The 21st-Century Engineer

Also of Interest

Changing Our World: True Stories of Women Engineers, by Sybil E. Hatch (ASCE Press, 2006). Real-life stories about the lives and careers of hundreds of women engineers, celebrating their contributions to every aspect of modern life. (ISBN 978-0-7844-0841-4)

Civil Engineering Practice in the Twenty-First Century: Knowledge and Skills for Design and Management, by Neil S. Grigg, Marvin E. Criswell, Darrell G. Fontane, and Thomas J. Siller (ASCE Press, 2001). Supplement to the technical preparation of engineers and other professionals, providing essential skills and strategies needed by civil engineers to be successful in the 21st century. (ISBN 978-0-7844-0526-0)

Degrees of Belief: Subjective Probability and Engineering Judgment, by Steven G. Vick (ASCE Press, 2002). Examination of the intersection of probability and risk analysis with professional judgment and expertise from a geotechnical perspective. (ISBN 978-0-7844-0598-7)

Effective Training for Civil Engineers, 2nd Edition, by H. Macdonald Steels (Thomas Telford, Ltd., 1999). Practical book that explores the most efficient methods of training for civil engineers. (ISBN 978-0-7277-2709-1)

Engineering Your Future, Second Edition, by Stuart G. Walesh (ASCE Press, 2000). Supplement to the technical preparation of engineers and other professionals, providing valuable advice and instruction crucial to success in today's world. (ISBN 978-0-7844-0489-8)

Managing and Leading: 52 Lessons Learned for Engineers, by Stuart G. Walesh (ASCE Press, 2004). Useful ideas and fundamental principles for engineers to improve their management and leadership skills. (ISBN 978-0-7844-0675-5)

The 21st-Century Engineer

A Proposal for Engineering Education Reform

PATRICIA D. GALLOWAY, PH.D., P.E.

Library of Congress Cataloging-in-Publication Data
Galloway, Patricia D.
  The 21st century engineer : a proposal for engineering education reform / Patricia D. Galloway.
    p. cm.
  Includes bibliographical references and index.
  ISBN 978-0-7844-0936-7
  1. Engineering--Study and teaching. 2. Engineers. I. Title.
  T65.G17 2007
  620.71'1--dc22

2007034432

Published by American Society of Civil Engineers
1801 Alexander Bell Drive
Reston, Virginia 20191
www.pubs.asce.org

This book is dedicated to. . .

. . .my husband, Kris R. Nielsen, who has stood by me in all that I have done;

. . .my mother, Maudine Frisby, who told me anything is possible if you put your mind to it; and

. . .ASCE, the backbone of our profession, which I am proud to have served as president.

"We have to fight this battle upon principle, and upon principle alone. . . . So I hope those with whom I am surrounded have principle enough to nerve themselves for the task, and leave nothing undone that can be fairly done to bring about the right result."

—Abraham Lincoln, July 17, 1858

Contents

Foreword

CHANGE IS A CONSTANT IN SOCIETY, AND ENGINEER-
ING helps to drive change and is driven by it. Although
change has been evident in every age, this book rightly
describes the unique circumstances in which the engineering profession finds itself at this particular moment
in history. Patricia Galloway brings to this book the special insights of an engineer who has been at the forefront
of the practice of civil engineering for her entire career.
She has served in many leadership roles, which provide
her with perspectives that others may lack.

During the past five years, several seminal
reports identified the special challenges facing engineering today. We live in a time when technology
pervades society, a time that should be ideal for engineering. However, engineering has evolved through
a process of steadily increasing specialization, which
gradually divided the profession into smaller component parts. It is hard to find a central voice to speak
for engineering as a whole. Simultaneously, developments that speed communications, such as the Internet, have quickly changed the playing field for everyone, engineers especially. Moreover, in only a decade,
the global economy has absorbed the three billion
people of India, China, and other developing nations.
Engineers in these nations have shown themselves
ready and able to take on—at much lower wages—

work that previously was done in the United States. The times call for new approaches if U.S. engineers are to be successful in the future.

The issues confronting engineers and the engineering profession are now largely agreed upon, but there is far less consensus about what engineers and the profession should do about those issues. Good ideas are emerging from meetings of engineering professionals and educators. It remains to formulate these into a coherent approach.

In this new book, Dr. Galloway offers a unified approach to help engineers prepare for the future in the face of uncertainty. She understands that we need to continue to educate young engineers in the fundamentals, but we must also do much more. The bottom line is that U.S. engineers must be well-rounded and have additional skills if they are to justify the higher wages they hope to command. She calls for an education that is inclusive of communication skills, ethics and professionalism, and leadership. She recognizes that engineering in the United States should build on its diverse population and use this as a strength. She also appreciates that, with these new skills, engineers can and should expand their role in public policy beyond design and manufacturing. As she points out, if engineers leave policy making to others, they will be unable to ensure that technology is used as a positive force in society.

Not averse to controversy, Dr. Galloway challenges engineering educators to move reform to the front burner. She demonstrates the limitations of the present curriculum, as well as those of the four-year degree. She issues a call to identify the master's degree as the first engineering professional degree, and she argues that it is past time to create a master's degree in professional engineering management. Under her proposal, such a degree would give engineers the skills needed to successfully lead engineering firms or large corporations, as well as to participate at the highest levels of government. Such a degree would also reinstate the value of a generalist—what might be termed the "master engineer"—who knows how to put all of the pieces together.

Not everyone in engineering practice or education will agree with every one of Dr. Galloway's views. Her book comes at a key moment, however, challenging us and stimulating us with ideas and a call to action. As she reminds us, time is not the engineer's friend. Dr. Galloway is to be congratulated for writing a much-needed book arriving at a critical moment. This book should be read and absorbed by those who want to make a difference.

G. WAYNE CLOUGH, PH.D., P.E.
President, Georgia Institute of Technology

Preface

I DECIDED TO WRITE THIS BOOK IN ORDER TO PROPOSE a framework for much-needed engineering education reform that can be used by all engineering disciplines. It has long been my view that engineers did not hold the leadership positions during the 20th century that engineers held during the 19th century and previously, and I believe the reason for this is an engineering educational system that has not kept pace with the demands of the marketplace. Conversations that I have had through the years with prominent individuals in major decision-making positions have convinced me that engineers are not perceived as effective communicators, as being sufficiently knowledgeable about the business side of the design and construction industry, and as possessing the "people" skills needed to assume leadership roles.

Engineering education reform is a global crisis that is being brought into sharper focus as the world begins to confront such transnational issues as, for example, climate change, decaying infrastructure, natural disasters, homeland security, widespread poverty in underdeveloped nations, an increasing demand for potable water—issues that cannot be solved by any one nation alone or any one profession alone. It is time for engineers to do something about this educational crisis and for universities to work in conjunction with industry and professional engineering organizations to devise solutions for better preparing the engineers of the future and for better equipping engineers currently working in professional practice. It is my hope that this book will generate the energy and enthusiasm that ultimately will bring about the change needed to secure the future of the engineering profession.

This book is founded on the concept that *all* engineers need to broaden their skill sets if they are to become leaders in business and public policy and if engineering is to be accorded the same status by the public that is accorded to the medical, legal, and accounting professions. While the views presented in the book are my own, they are based on the research I conducted in pursuit of my doctorate; my 30 years of experience in the engineering and construction industry; my leadership positions in professional organizations, among them the Project Management Institute, the Pan American Academy of Engineering, the National Academy of Construction, the Society of Women Engineers, and the American Society of Civil Engineers—serving as the first woman president in the 152-year history of the latter Society; my service on various advisory boards and committees, including the Purdue University engineering dean's faculty advisory committee and the engineering and international advisory boards of the National Science Foundation; the National Science Board; and my travels to nearly one hundred countries during the course of writing this book. It is my intent that this book bring engineers together—not only here in the United States, but across the globe—and that it elevate the debate on engineering education reform to a more informed level so that very substantial progress is made within the next several years.

The book's audience is not limited to engineers in the United States. Rather, the book is geared toward engineers the world over interested in enhancing their skill sets, be they company executives who are trying to determine what continuing education is needed for their employees so their companies will succeed in the 21st century; students who want to enhance their studies and get a head start in their careers; officials with government entities—both in the United States and elsewhere—as they face critical engineering issues and determine how best to solve them; and educators, who can truly make a difference in engineering education reform.

By the time you finish reading this book, you will have a better understanding of what a 21st-century engineer may be facing in his or her endeavors, of how we as a profession must elevate our public standing and increase the public's confidence in us, and of what skills may be necessary to add to one's individual tool box in order to succeed in the 21st-century global marketplace. Perhaps most important of all, the book provides a model for engineering education reform that was inspired by the curriculum being explored at Kochi University of Technology in Japan, where I earned my doctorate in infrastructure systems engineering.

Preparation of this book would not have been possible without the efforts and support of certain individuals, and I would be remiss if I did not recognize them. First and foremost are my mother, Maudine Frisby, and my husband, Kris Nielsen. My mother inspired and motivated me throughout my childhood and instilled within me the importance of education and the need to pursue education on a lifelong basis. She told me that nothing is impossible and advised me never to believe anyone who tells me that I cannot accomplish something. It was she who introduced me to the three Cs—communication, confidence, and commitment—which taken together constitute a foundation of leadership that I discuss in chapter five. It was her own confidence in me and her personal commitment that enabled me to set my own goals and work toward achieving those goals, and it was her constant clear communication to me that gave me a sound foundation upon which to build my life's work.

My husband—my best friend in many ways—gave up time in his own career to support me in my career. He is the wind beneath my wings and has helped me to achieve much of what I have achieved. He was and continues to be my mentor, and he possesses a wealth of knowledge that I find unmatched in others. His insights have always proven invaluable to me.

I wish to thank Dr. Sunji Kusayanagi, my adviser at Kochi University of Technology. Without his insistence that I go back to school to earn my doctorate, I doubt that I would have been able to write this book, which is largely based on my dissertation.

I also wish to offer my sincere appreciation and thanks to all of the officers of my firm, The Nielsen-Wurster Group, Inc., who have put up with my ranting and raving on this subject for years and who have actively supported my professional activities, including my year of service as the president of the American Society of Civil Engineers.

Patricia D. Galloway

Introduction

THE WORLD CHANGED MORE DURING THE PAST HUN-
dred years than during any preceding century. By the
dawn of the new millennium, the developed world
constituted the healthiest, safest, and most productive civilization in history—a civilization that has
advanced and continues to advance with extraordinary rapidity, in large measure through the achievements of the engineering profession. But the speed
with which this civilization continues to advance is
fundamentally challenging the way in which engineering is practiced and the way in which engineering
students are educated. Technological breakthroughs—
the Internet most prominent among them—have
effected an increasingly global "workplace" in which
the collaborative efforts of multinational teams are
unhampered by geographical distance or time zones,
and this global workplace in turn is introducing new
imperatives not only to engineering practice but to
engineering education as well.

    If engineers are to compete successfully in this
global workplace and establish themselves as leaders in solving many of the world's most pressing
problems, they must embrace the need for professional innovation and they must do so quickly. They
must understand that long-established methods of

practicing engineering and educating engineers are in critical need of reform, and they must understand that they must act decisively within the next several years to ensure that these reforms are adequately formulated and implemented.

Engineers must also embrace the need to enhance their image on the world stage—to elevate their professional standing to the highest level. Engineers by and large have come to be viewed as technicians—as commodities of sorts, not as practitioners of a profession engaged in a high calling comparable to the practice of law or medicine, for example. It is essential that engineers alter this perception. Of course, in order to elevate their professional standing on the global stage, engineers must understand why and how they must do so.

Central to the question of *why* they must do so is the engineering profession's very survival. If engineers are relegated to the role of technician, they will no longer command the levels of responsibility that will enable them to successfully compete in the global economy or assume the leadership roles that will enable them to elevate standards of living worldwide and provide enhanced protection of the environment. Central to the question of *how* they must do so is engineering education reform. Engineers must understand that the four-year engineering degree is no longer adequate. The engineering student will not—indeed cannot—learn within a four-year time frame everything that he or she will need to know in order to perform adequately, let alone successfully, as an engineering professional in this rapidly advancing global marketplace. Once they become practicing professionals, engineers must understand that the world in which they are working is an ever-changing milieu: not even the fundamentals are fixed as new technologies continue to emerge. Engineers must accept full responsibility for their own continued education, and engineering schools must prepare them to do so by teaching them how to learn on a lifelong basis.

Today's engineers simply do not possess many of the proficiencies needed to compete internationally or to advance the engineering profession. While engineers remain strong in terms of their technological skills, they are generally weak in terms of their management and communication capabilities. Moreover, they do not fully understand the concept of globalization; they do not have a firm grasp of the issues confronting the 21st-century engineer; they lack the competencies that would enable them to rise to leadership positions within government and industry; and they are not developing curricula that would train engineers to anticipate and focus on the rapid changes by which the 21st century will be at least partially defined.

A solid understanding of globalization is key to an engineer's success in today's global society. Globalization involves the ability to understand that the world economy has become tightly linked with much of the change triggered by technology; to understand other cultures, especially the societal elements of these cultures; to work effectively in multinational teams; to communicate effectively—both orally and in writing—in the international business language of English; to recognize and understand issues of sustainability; to understand the importance of transparency while working with local populations; and to understand public policy issues around the world and in the country in which one is working. It will be these fundamental capacities that will enable 21st-century engineers to develop into professionals capable of working successfully both domestically and globally, highly respected by the general public and regarded by governments the world over as professionals of the highest order.

The reason why it is so imperative that engineers fully understand and embrace the concept of globalization is that many of the complex issues of the 21st century can be addressed only through engineering collaboration between nations. These issues include the increase in aging populations and the escalating health care costs associated with these aging populations; decaying infrastructure; the increasing demand for potable water; responsible consumption and protection of natural resources; homeland security and public safety; global warming; natural disasters; and ethics, bribery, and corruption in the global workplace.

The question of whether engineers possess the skills necessary to ensure that their profession will endure has been debated for decades. While the early pioneers of engineering had a clear understanding of the skills they needed in order to thrive as a profession, many of today's engineers do not. Present-day engineers are seemingly encased in the past, believing—mistakenly—that technological prowess is all that is needed to succeed. Although engineers have become better and better at mastering the technical skills associated with emerging technologies and inventions that are being developed at ever-increasing rates, they have little or no training in the "soft" skills required to succeed in today's global professional community. Although engineering is still a respected profession, the professional standing of the engineer has diminished over the years, and this in turn has resulted in lower remuneration than that enjoyed, for example, by practitioners of law or medicine.

In its 2004 book *The Engineer of 2020: Visions of Engineering in the New Century,* the National Academy of Engineering defined the engineer of the 21st century as follows:

> Engineering is problem recognition, formulation, and solution. In the next twenty years, engineers and engineering students will be required to use new tools and apply ever-increasing knowledge in expanding engineering disciplines, all while considering societal repercussions and constraints within a complex landscape of old and new ideas. They will be working with diverse teams of engineers and nonengineers to formulate solutions to yet unknown problems. . . . The engineers of 2020 will be actively involved in political and community arenas. They will understand workforce constraints, and they will recognize education and training requirements necessary for dealing with customers and the broader public. Engineering will need to expand its reach and thought patterns and political influence if it is to fulfill its potential to help create a better world for our children and grandchildren. (NAE 2004, 43–44)

The practice of engineering continues to grow increasingly more complex. As a result of the rapid rise of information technology, the explosion of knowledge in engineering and construction, the enhanced public awareness of and involvement in engineered projects, and the growing complexity of civil infrastructure systems around the world, the job performed by the engineer continues to become more demanding. This trend is almost certain to accelerate, and as a consequence engineers will be expected to possess both a breadth of capability and a specialized technical and managerial competence that are greater than was required of previous generations. These requirements will demand that the engineering profession acquire a new skill set in order to perform successfully and be respected by the public. Engineers must become more aware of the need to work in teams, consider social issues, understand political and economic relations between nations and their peoples, and understand intellectual property, project management, multilingual influences, and cultural diversity, as these factors will drive the engineering practice of the 21st century. It will become essential that engineers know how and when to incorporate social elements into a comprehensive systems analysis of their work.

One of the primary reasons why the image of the engineering profession has become diminished in the view of the general public is that during the 20th century engineering splintered to include a greater number of diverse activities, including telecommunication and information technology. Engineering has had to embrace a number of other professions that had not previously been associated with engineering, and the engineering world has become more multidisciplinary. In fact, technology, created through the process of engineering, is now recognized by governments and corporations around the world as the driving force for profitable growth and sustained societies. The role that engineers play in society and the benefits that society reaps from the work of engineers are not evident to the public, which tends to view engineering as a tedious and uncreative profession associated with the economy of the past. The meaning of the term "engineering" has become lost somewhere between science and technology. Thus, there is an immediate need for engineers to obtain the skills necessary to better communicate with the public and to assume the lead on projects.

Clearly, the need to educate the engineer of the 21st century differently—or more precisely, more strategically—is essential to the endurance of the profession. In 1986, the National Research Council's Panel on Undergraduate Education published a report that cited three "serious concerns" with respect to engineering graduates: First, many have "little knowledge" of the design process; second, they have "inadequate knowledge of the role of technology in their professions"; and third, they have "little knowledge of business, economics, and management" (NRC 1986).

In *Transnational Competence: Rethinking the U.S.–Japan Educational Relationship,* John Hawkins and William Cummings focus specifically on the interactions between the United States and Japan and explain the principles and concepts behind increased internationalization. The concept of transnational competence is noted to be the "thoughtful integration of technical and cultural skills," with six major components:

1. It is the ability to imagine, analyze and creatively address the potential of other local economies and cultures.
2. It is a knowledge of commercial, technical, and cultural developments in these other locales.

3. It is an awareness of who the key lenders of these locales are (and the ability to engage them in useful dialogue).
4. It is an understanding of local customs and negotiating strategies.
5. It is having skills in business, law, public affairs, or technology.
6. It is having a facility in English, at least one other major language, and computers. (Hawkins and Cummings 2000, 8)

The American Society of Civil Engineers (ASCE) has long recognized the need for engineering education reform. In October 2001, the society established the Task Committee on Academic Prerequisites for Professional Practice and asked the committee to "develop, organize, and execute a detailed plan for full realization of ASCE Policy Statement 465—'Academic Prerequisites for Licensure and Professional Practice.'" In November 2001, the society's Board of Direction adopted ASCE Policy 465, which "supports the concept of the master's degree or equivalent as a prerequisite for licensure and the practice of civil engineering at the professional level" (ASCE 2001). This policy statement led to the creation of a formal body of knowledge for civil engineering. In 2007, the ASCE board approved an updated Policy Statement 465, which incorporates refinements that reflect progress and change within the profession.

At a press conference held at the National Academy of Engineering in Washington, D.C., on February 25, 2004, ASCE presented its report *Civil Engineering Body of Knowledge for the 21st Century: Preparing the Civil Engineer for the Future* (ASCE 2004) and, in essence, took a stand on the requirements that the 21st-century engineer must fulfill in order to become a licensed professional:

> The first proposal by any engineering discipline to take the educational prerequisites for engineers beyond the four years of formal education, the report also calls for the acquisition of experience before a civil engineer can sit for the licensure exam. It may also be possible for distance learning programs and corporate and government agency educational programs to substitute for traditional academic paths beyond the baccalaureate. . . . The Society hopes that the report will lead to a revision of current undergraduate and graduate programs to reflect the basic skills, knowledge, and attitudes that will be expected of professional civil engineers and that eventually new programs will be developed. (*ASCE News* 2004, 1)

ASCE continues its work on the body of knowledge and, as of the publishing of this book, has released a draft of the second edition (ASCE 2007b). The key differences between the first and second editions focus on the aspirational vision for civil engineering, an increase in the number of outcomes, and a more highly structured approach to the level of achievement expected for each of the outcomes, at various stages of a civil engineer's education. The final version of this report is due in February 2008.

As ASCE began to set the foundation for "raising the bar," other professional societies including the National Society of Professional Engineers and the National Academy of Engineering began to take note and offer support. The National Council of Examiners for Engineering and Surveying (NCEES) voted in August 2005 to change the *Model Law* for professional engineer licensure to include additional education beyond the bachelor's degree beginning no earlier than 2010 (ASCE 2005). In 2006, NCEES added language to its Model Law to include education beyond the bachelor's degree (see NCEES 2007).

The body of knowledge is being prepared to establish the knowledge, skills, and attitudes necessary for engineering graduates to effectively enter licensed practice. ASCE's first report built upon the existing eleven "outcomes" identified by the Accreditation Board for Engineering and Technology (ABET 2006), by adding four additional outcomes to those incorporated within existing accredited U.S. undergraduate curricula. The additional outcomes include additional technical engineering depth in one or more areas and additional breadth in the areas of project management, construction, and asset management; business and public policy and administration; and leadership. ASCE's second body of knowledge report will propose additional outcomes, which as of this publication are under review. The additional breadth components might eventually be incorporated into undergraduate engineering curricula, leaving additional education beyond the bachelor's degree to consist of a flexible program of additional depth and breadth in engineering and professional topic areas.

Similar needs are being recognized in other parts of the world. For instance, as outlined in the Asia-Pacific Economic Cooperation *Engineer Manual*, the organization considers it important for its member engineers to possess skills in general management, project management, quality assurance and total quality management, marketing of engineering products or services, financial or human resource management, the design and delivery of training programs, policy development, and regulation development. The manual also notes that these activities will typically involve leadership, teamwork, oral and written communication, presentations, and interpersonal skills in the practice of all engineering disciplines (APEC 2000).

Such skill assessments were discussed in 2004 at the Society of Petroleum Engineers' (SPE) sixth colloquium on petroleum engineering education. The three-day roundtable was attended by representatives of academia, industry, and the service sector to discuss how universities—in partnership with industry, service companies, and government—could help enhance engineering education. The objective was to determine how the world could come together on the establishment and global acceptance of certification standards, innovative teaching programs, and agreement as to the skills and attitudes required for the "new engineering professional." The SPE skills matrix has undergone changes to reflect this input. A 2006 survey of more than 800 licensed engineers confirmed the breakdown of management expectations and competencies and, thus, the critical need to address the current deficiencies in engineering education (SPE 2006). The skills being sought include effective written and oral communication skills, the ability to work well on a team, analytical and problem-solving skills, judgment and decision-making capabilities, organizational and planning skills, information technology skills, an understanding of ethics, and an understanding of the business and legal aspects of professional practice. The goal is for universities to develop more holistic curricula.

At present no engineering program—either at the undergraduate or graduate level—includes instruction in all of the skills mentioned above. Thus it is essential that engineering education be reformed. As noted by the National Academy of Engineering,

> In the past, changes in the engineering profession and engineering education have followed changes in technology and society. Disciplines were added and curricula were created to meet the critical challenges in society and to provide the workforce required to integrate new developments in our economy. Today's landscape is little different; society continually changes and engineering must adapt to remain relevant. . . .
>
> While certain basics of engineering will not change, the global economy and the way engineers will work will reflect an ongoing evolution that began to gain momentum a decade ago.

The economy in which we will work will be strongly influenced by the global marketplace for engineering services, a growing need for interdisciplinary and system-based approaches, demands for customerization, and an increasingly diverse talent pool. The steady integration of technology in our infrastructure and lives calls for more involvement by engineers in the setting of public policy and in participation in the civic arena. The external forces in society, the economy, and the professional environment pose imperatives for change that may exceed those to come from the changes expected in the technology engineers will have at their disposal in 2020. . . . If the engineering profession is to take the initiative for its future, it must 1) agree on an exciting vision for its future; 2) transform engineering education to help achieve the vision; 3) build a clear image of the new roles for engineers, including broad-based technology leaders, in the mind of the public and prospective students who can replenish and improve the talent base of an aging engineering workforce; 4) accommodate innovative developments from nonengineering fields; and 5) find ways to focus energies of the different disciplines of engineering toward a common goal. (NAE 2004, 1–5)

While endeavors are under way in the United States, the United Kingdom, and Asia to revamp systems for the certification of engineers and while progress is being made in the accreditation of universities to ensure consistent engineering programs in Asia—particularly in Japan—there is currently no adequate educational system for the engineer of the 21st century in any developed nation. No engineering educational program at either the undergraduate or graduate level teaches the fundamental capacities now required. Developing a proposed education program through modifications of successful programs to meet the needs of the 21st-century engineer will enable engineers to acquire the skills necessary to succeed in the world economy and to grow within the domestic construction market in a more transparent manner.

Research has demonstrated that the reason for the declining image of the engineering profession stems from an engineering educational system that is inadequate—specifically, that lacks the management and communication courses that impart a critical set of skills needed to succeed in the global marketplace. In this book, I discuss the skills necessary to survive and succeed in today's global marketplace.

I also propose and provide a framework for a master's program in professional engineering and management.

If engineers are to remain competitive in the global marketplace, universities simply must reform their engineering curricula. The time for taking the initiative to ensure the future of the engineering profession has come.

1

Globalization

ONE OF THE FUNDAMENTAL CONCEPTS THAT THE engineer must fully understand if he or she is to succeed in today's global society is globalization. The term *globalization* is used to describe the changes in societies and the world economy that are the result of dramatically increased trade and cultural exchange. Globalization refers to the increasing integration of nations through tariffs, investment, transfer of technology, and the exchange of ideas and cultures. As defined by Globalization101.org, an Internet resource developed by the Carnegie Endowment for International Peace, "globalization is a process of interaction and integration among the people, companies, and governments of different nations, a process driven by international trade and investment and aided by international technology. This process has effects on the environment, on culture, on political systems, on economic development and prosperity, and on human physical well-being in societies around the world" (CEIP 2007). The process is characterized by an increase in the connectivity and interdependence of the world's markets and businesses. Ten years ago, the entire market economy population of the world was composed of approximately 500 million to 600 million people residing primarily in the United

States, Canada, Japan, and Western Europe. Today, however, more than six billion people reside in the world market economy, which has expanded to encompass much of the globe.

Immediately following World War II, globalization came to be identified with greater international movement of commodities, money, information, and people and the development of technology, organizations, legal systems, and infrastructure to facilitate this movement. In more recent years, globalization has been associated with an increase in the global population; an increase in international trade at a faster rate than the growth of the world economy; an increase in the international flow of capital, including foreign investment; greater transborder data flow by means of such technologies as the Internet, communication satellites, and telephones; greater international cultural exchange; the expansion of multiculturalism; an erosion of national sovereignty and national borders by means of international agreements leading to the establishment of such entities as the World Trade Organization; the development of global telecommunications infrastructure; the development of global financial systems; an increase in the share of the world economy controlled by multinational corporations; an increased role by such organizations as the World Trade Organization, the World Intellectual Property Organization, and the International Monetary Fund, which deal with international transactions; and an increase in the number of standards applied globally—for example, copyright laws and intellectual property rights.

The global population's explosive growth in particular will exert a tremendous impact on the health of the environment, the maintenance of infrastructure, the availability of potable water, waste management, and other significant health and safety concerns—on issues, in other words, in which the engineering profession will be closely involved. Over the course of the last century, the world's population burgeoned from 1.6 billion to 6.7 billion, and researchers estimate that within the next 10 to 15 years it will increase by yet another billion. In fact, so great is this growth rate that the Population Division of the Department of Economic and Social Affairs of the United Nations projects that the world's population will reach 8.9 billion by 2050 and will continue to increase well into the 22nd century (United Nations 2007).

In the 21st century, advances in engineering will, to a large measure, determine the rate of world economic growth, the quality of life, and standards of health and security. The practice of engineering will become increasingly important in addressing many critical global issues. Engineering has truly become a global enterprise as new ideas and discoveries have emerged all over the world and the balance of engineering expertise has shifted among various countries. Engineering advances will increasingly depend on the ability to draw upon the best minds regardless of their national origins. Thus, as engineers embark on the new millennium, they must remain acutely aware of such issues as climate change, human practices that are destructive to the environment, the devastation wrought by natural disasters, aging infrastructure, development of critical infrastructure in light of the threat of terrorism, and energy and water shortages—all of which threaten global sustainability as well as global stability. Engineers have the opportunity to develop innovative solutions to these global problems—to elevate their professional standing on the global stage and enhance their image in the public's eye. And while this will require engineers to employ the most up-to-date relevant technology, it will also require them to communicate effectively to ensure the accurate and efficient transfer of data across international borders.

Globalization, therefore, is a concept that engineers must not only understand but incorporate into their everyday business and professional lives. The mission of the 21st-century engineer is to ensure that there are no boundaries in terms of how engineering is applied to better the quality of life worldwide. But in order to do so, the 21st-century engineer must be better prepared. The 21st-century engineer must be facile in adapting to changes in global forces and trends to ethically assist the world in establishing a balance in the standard of living for developing and developed nations alike. Deftly adapting to these changes in global forces and trends will determine how well the 21st-century engineer succeeds on the new global business landscape.

The New Business Landscape

The world's economy has become tightly linked with recent rapid advances in technology—in particular with the Internet, which can be credited as the real leveler of the playing field within the global marketplace. For the Internet has given rise to e-commerce and to the 24/7 global workplace that operates 24 hours a day, seven days a week. And engineers, therefore, must now think and plan strategically on this new global business landscape on which the workday in essence never comes to a close. But the ability to function successfully on this landscape entails much more than a faculty to use the Internet to communicate and transfer information 24/7. It also entails a grasp of the fact that success on this landscape requires the effective management of knowledge technology. The business acumen that comes from a command of information technology is going to be required to succeed in the 21st century.

As a result of globalization, more and more businesses are pursuing the very same customers. Customers, in turn, have become more sophisticated and are better informed buyers. Information technology enables these customers to quickly locate and analyze products and services and to make more intelligent choices. The 21st-century global society is an increasingly knowledge-based society, and knowledge and continuous learning are now critical elements of success within this society. On this new business landscape, the knowledge component of services has expanded dramatically in terms of importance and has become the dominant component of customer value. Those who will succeed on this new landscape are those who manage knowledge technology effectively—or more precisely, better and more quickly than their competitors. In short, engineers must think and plan strategically in the world economy, as opposed to confining their thinking to their own domestic economy. Engineers must be sophisticated knowledge managers and strategists who understand the needs of the global society and who assume leadership roles in devising means of satisfying those needs.

One of the other challenges confronting the engineer of the 21st century is how to best manage today's megaprojects. Ten years ago, projects on the scale of China's Three Gorges Dam, Boston's Big Dig, Dubai's offshore human-made islands, the Channel Tunnel Rail Link, the widening of the Panama Canal, or London's Crossrail project would have seemed unimaginable. Today, however, they represent a new breed of large projects that involve multinational firms in their design and construction and cost billions of dollars to complete over the course of several years. In the 1970s and 1980s, the World Bank voiced concern about how these megaprojects would affect the construction industry. These megaprojects are absorbing enormous amounts of capital over extended periods of time, and their complexity poses significant challenges in terms of their management. Given the fact that such projects take 12 or 15 or 20 years to complete, how many people today can claim to have managed more than one megaproject? In the 21st century, engineers must become comfortable managing megaprojects worldwide, and in order to do so successfully they must understand globalization and the new business landscape effected by globalization.

For U.S. engineers practicing in small firms—with, say, fewer than 35 employees—megaprojects and globalization might seem only distantly relevant. If, however, these engineers look around, they will see that some of the emerging technologies and innovative ideas are being generated by engineering colleagues from all around the world. Being small and local does not preclude a responsible engineer from maintaining competency and knowledge of the most effective and efficient ways to solve a problem. Value can be gained from learning what others do around the globe.

Globalization comes with related issues, such as the offshoring of engineering jobs because of lower-cost labor in some countries and the way the Internet encourages a global workforce. Countries such as India and China are quickly reforming their engineering curriculums to match those in developed nations. However, their wages remain 40 to 80 percent lower. Offshoring affects not only job displacement but it directs job growth to lower-cost nations (Kenney and Dossani 2005). This, in turn, could have a direct impact on engineers' salaries—no matter whether they are in a small local firm or a huge multinational one. The real downside emerges as it becomes difficult to attract young people to the engineering profession, especially at a time when engineers are desperately needed to address failing infrastructure, climate change, and natural disasters.

Thus, as the global landscape changes, so must the skill sets of engineers if they are to resist the pressures of low wages and technical/electronic mobility. Understanding globalization is a key component of the engineer's toolkit in the continuous quest to raise the profession's image and hold leadership positions—regardless of location or size.

Sustainability

There is perhaps no greater need on earth at this moment than sustainability. Distilled into its simplest form, sustainability is the practice of adequately meeting current needs while ensuring that future needs will be adequately met. Fleshed out a bit more, sustainability is the practice of ensuring that all of the world's inhabitants—from those living in the most developed nations to those living in the most underdeveloped nations—are ensured adequate food, shelter, and sanitation, now and in the future.

One of the greatest challenges to ensuring sustainability is population growth. The world's population is expanding at the rate of between 80 million and 100 million per year and this growth shows no signs of abatement. The three segments of the world's population— those living in developed nations, those living in developing nations, and those living in underdeveloped nations—are not equal in terms of need, however. The needs of each sector are quite different, and the engineer must understand those differences and how best to address them when planning infrastructure projects that adhere to the principles of sustainable design.

Within the developed nations, which are the most advanced, urbanization presents the greatest challenge. For example, nearly 80 percent of the population of North America resides in urban areas despite the vast areas of land available to support this population. Comparable percentages of urban dwellers are found in Asia, Europe, and Australia as well. In the developed nations, therefore, the 21st-century engineer must address an aging infrastructure, which was constructed at a time when little consideration was given to sustainable development. Major urban areas are confronting problems associated with decaying infrastructure systems, air pollution, traffic congestion, the destruction or despoliation of such natural resources as trees and streams, and the contamination of drinking water sources. The engineer's challenges here focus on repairing and rehabilitating existing infrastructure to meet the needs of future generations by providing more sustainable infrastructure and on devising methods of mitigating air pollution, lowering carbon emissions, providing clean water, and transporting people by means other than gasoline-consuming vehicles. With respect to new construction, the design and construction of green buildings pose for engineers considerable new challenges that include the need for additional education, conformance to new standards, and collaboration with multiple stakeholders—the latter a process with which many 20th-century engineers were not entirely comfortable.

The challenges in developing nations are concentrated on improving infrastructure to support rapidly growing urban populations. Growing urban populations, however, pose challenges not only within developing nations but globally as well. In 2000, 47 percent of the world's population resided in urban areas. The United Nations estimates that by 2030 the urbanization rate will be more than 60 percent (United Nations 2004). While this urbanization will greatly promote social and economic development, it will also create social, economic, and environmental problems. The models used in the past in the developed nations, however—models that inflicted environmental abuses, for example—must be abandoned in favor of new models that emphasize sustainability. Lessons learned from engineering and construction projects in urban areas developed in the past must be carefully reviewed, and a concerted effort must be made not to repeat the mistakes of the past.

Underdeveloped nations might at first appear to present the greatest challenges. However, the infrastructure requirements of an underdeveloped country are quite basic: meet the essential needs of human life—that is, adequate food, water, and sanitation. Establishing the ability to transport agricultural products to market, to food-processing plants, and then to the population and the ability to construct safe and reliable water and sanitation systems are the most pressing priorities. Engineers can play a role in determining what types of food-processing plants are required in a region as well as in determining what means of transport would be most efficient in delivering the food to the population. And while sanitation is key to preventing disease among the populations of underdeveloped nations, thoughtful consideration must be given to the level of sophistication of any proposed sanitation plant or water treatment process. For example, advanced wastewater plants may seem an obvious solution. But if a location is so remote as to make the delivery of spare parts difficult or if the educational level of a population is such that maintenance cannot be performed satisfactorily, then an advanced wastewater treatment plant may not be the best approach. Perhaps a gravity-fed sewer system would be the most practical solution. In any event, the engineer must be capable of recognizing potential problems and devising the most practical solutions for ensuring sustainability.

An important point to bear in mind is that the level of technological advancement in developing and underdeveloped nations significantly defines the parameters within which the 21st-century engineer will be working in these nations, and it is important for the engineer to understand these technological limitations as well as the potential for future technological advancements. In a 2006 report, the RAND Corporation raises the specter of a growing divide between those nations that are technologically advanced and those nations that are not. As the report notes, different countries use technology applications to solve problems different ways, and some countries will not be prepared in 15 years to fully use the least demanding of these applications even if they should acquire them (Silberglitt et al. 2006).

The questions the 21st-century engineer must answer before embarking on any project in a developing or underdeveloped nation are these: Is the current population able to use and maintain the technologies to be employed on a project? Does the population have the ability to deal with the required computer and other hardware upgrades that must be performed?

Global Infrastructure Risk

Another important area of which the 21st-century engineer must become knowledgeable in order to succeed in the global marketplace is the understanding of what risks face projects today and how to identify, assess, and monitor those risks, including the financial aspects. The crisis in the emerging markets in the 1990s made it quite clear that the opportunities of globalization do not come without risk—risk arising from volatile capital movements and risk of social, economic, and environmental degradation created by poverty. Engineers have not typically been viewed as capable of managing the financial aspects of projects. However, in today's global marketplace, understanding the financial aspects is fundamental to serving as the leader of a project team. Investment and financing decisions regarding potential projects around the world are often made on the basis of limited analysis of the project execution and the project's future use or operation. Then financial viability is assessed, assuming little risk to achieving scope, function, timing, and cost goals. Such an approach leads to a highly inefficient use of capital.

The potential for such inefficiency is increasing. For example, recovery from the Asian downturn that began in 1996 is still limited. This downturn led to regional political crises, instability, and financial retrenchment in many geographical areas, especially in Asia and the Western Pacific. Thus, easy financing in this geographical area is a thing of the past. Similarly, the cooling and recovery (albeit in its infancy) of the U.S. economy leads to more thorough potential project risk review globally before project financing is forthcoming. The

war on terrorism reprioritizes near- and medium-term project requirements and shifts public sector project spending focuses worldwide. All
business sectors and their capital project needs are forced to adjust to
this new commercial reality.

Nonetheless, global needs continue to increase dramatically.
Available capital has shrunk, and there has been only recent minimal
movement toward a growth trend. There is now significant competition
for available financing reminiscent of the tight markets of the 1980s.
Project risk is a necessary focus that is driving global and regional project financing decisions, and it is a concept that must be understood by
today's global engineer. Significantly, medium- and long-term market
demand for the products/commodities in the process industries and
resource production sectors is indeed shifting away from yesterday's
models. Public project spending will not lead directly, and in the near
term, to industrial demand. For example, public spending on security
is expensive, as is public spending to ensure political stability. Such
expenditures often require infrastructure development and evolution.
This form of infrastructure does not improve productive efficiency;
it merely provides a safer and more stable environment in which to
live. In the past, airport capital projects would expand capacity, use
efficiency, and transportation timeliness—all of which translate into
immediate business expansion. Conversely, today's plant and transportation infrastructure security will not translate into immediate
business benefit, but rather will provide longer-term business environment risk reduction.

In light of the disparate needs of populations around the globe,
the engineer must recognize that these various projects will be constructed in the midst of various risks. As the project leader, the engineer
must not only recognize what these risks are, but also how to identify, address, and provide the framework for solutions to best manage
them. Risks attendant to global projects—especially in developing and
underdeveloped nations—are not easily identified at first glance.

Risk within this context is not typical risk that can be insured.
It consists of those issues, conditions, and actions that prevent a project or project participant from meeting goals—for example, goals
of scope, quality, function, delivery time, and cost. Addressing risk
and its potential manifestation, its probable impact, and its management prospects is fundamental to successful engineering practice. Kris
Nielsen defines the two types of risks that should be assessed by engineers. The first group consists of project-specific risks:

- *Delivery/Operations.* This risk factor involves those issues or concerns associated with engineering, procurement, construction (EPC) execution and operation of the project.
- *Technology.* This risk factor involves those issues or concerns associated with the technologies involved in the EPC methods and operation technology of the project.
- *Financial.* This risk factor involves those issues or concerns associated with the financing of the project, including the EPC period and operations or equity financing.
- *Procurement/Contractual.* This risk factor involves those issues or concerns associated with the contractual and procurement approaches/systems/processes used for both EPC and operation of the project.

The second group consists of project context risks:

- *Political.* This risk factor involves those issues or concerns associated with the local, regional, and national political situation confronting the project.
- *Environmental.* This risk factor involves those issues or concerns associated with the environmental problems, concerns, and activities confronting the project during the EPC execution and the project operation.
- *Social.* This risk factor involves those issues or concerns associated with the social and cultural impacts of the project to the community and region within which it is to be located.
- *Economic.* This risk factor involves those issues or concerns associated with the macroeconomic impact of the project to the community and region within which it is to be located. (Nielsen 2006, 63–64)

Global public financing agencies, private banking and financing corporations, and government and corporate senior management now demand greater risk consideration. Projects do not move from feasibility to financing or funding without broad risk assessment. Projects deemed to squander capital are being canceled or refocused to reduce risk before any feasibility studies are conducted. When projects are refocused before feasibility studies are undertaken, the feasibility studies are subsequently used to reduce or eliminate risk before financing or funding is sought. The financial community or management/board of directors approving committees are now requiring risk evaluations by those seeking the financing and those internally considering the financing. The goal: improved and practical efficiency in using capital and in executing and operating projects with reduced risk.

Today, complex proprietary risk rating models are part of the solicitation process or the evaluation process required for project financing or funding. These models and the input upon which they operate ensure consistency among project evaluations, provide relative comparison with similar or competing capital uses, and tie to the financing or funding uses. Merchant and investment banks, public finance agencies and affiliates, and executive management entities are requiring assessments, evaluations, and ratings as part of proposals and applications.

Ultimately, worthy ideas, good intentions, and good planning do not bring a project to fruition—funding does. And in order to secure funding or financing, risk must be evaluated very thoroughly and rated within the categories of project-specific risks and context-specific risks. These risk categories focus on both initial project execution and ultimate project operations and are then remodeled to produce ratings that are consistent and comparable. In turn, the ratings underpin decisions on corporate funding versus other uses of capital. In the case of financed projects, the ratings underpin project financing decisions and the cost thereof to the party seeking it. As an example, the use of capital that meets a government's agenda the most effectively or that provides the corporate owner/operator with the better return generally receives the higher rating as less risk is involved. This means that there will be a less probable manifestation of risks that will erode the project goals. Capital funding or financing rates are thus lower.

Such risk assessment and risk management is an essential skill of the 21st-century engineer.

Cross-Border Mobility of Engineers through Licensure

While the global engineer must acquire a broad set of skills and be sensitive to myriad national and regional customs, practices, public policy issues, and politics, an international system for standardizing qualifications is essential to ensuring engineering quality worldwide.

It is clear that as the engineering profession has become more global, an internationally acceptable licensing procedure for the professional engineer is needed. In an effort to recognize the validity of engineering degrees obtained in various countries and to enhance the ability of engineers to practice globally, the Washington Accord—an international accreditation agreement for professional engineering degree programs—was signed in 1989 in Washington, D.C. (Washington Accord 2007). The goal of the accord is the mutual recognition of accreditation systems and professional engineering qualifications among the participating nations of the United States, Canada, Australia, Hong Kong, the Republic of Ireland, New Zealand, South Africa, the United Kingdom, and Japan. The accord does not, however, address the international licensing or registration of professional engineers.

Several years following the signing of the accord, representatives of the signatories—in conjunction with representatives of the European Federation of National Engineering Associations—met in March 1996, and they met with the Association of Japanese Consulting Engineers in 1997. Later in 1997, the Washington Accord signatories agreed to establish a forum to be known as the Engineers Mobility Forum through which they would develop, monitor, maintain, and promote mutually acceptable standards and criteria for facilitating the cross-border mobility of experienced professional engineers; seek to gain a greater understanding of the existing barriers to mobility and to develop and promote strategies to help governments and licensing authorities manage these barriers in an effective and nondiscriminatory manner; encourage the relevant governments and licensing authorities to adopt and implement mutual mobility procedures consistent with the standards and practices recommended by the signatories to such agreements as may be established by and through the Engineers Mobility Forum; and identify and encourage the implementation of best practices for the preparation and assessment of engineers intending to practice at the professional level. Additionally the Asia-Pacific Economic Cooperation Engineer Register was initiated in 2000 by the cooperative's economic leaders as part of initiatives to liberalize trade with regard to professional services and to ease the process for engineers to gain access to work and to practice in other countries.

The groundwork has been laid, then, for the international mobility of professional engineers, who—as the new business landscape becomes increasingly global—must seize the opportunity to play a vital role in addressing such critical concerns of the global community as ensuring sustainable development in the midst of explosive population growth; aging infrastructure; rapidly increasing urbanization; and technological inequities among developed, developing, and underdeveloped nations. In the 21st century, science and technology will become indispensable to the well-being of humankind and planet Earth, and engineers should be at the forefront of the development and application of much of this science and technology.

2

Communication

ASIDE FROM THE LACK OF UNDERSTANDING OF GLO-
balization, the number one "soft skill" that has
almost certainly kept many engineers from holding
leadership and political positions is their unwillingness or inability to communicate effectively. Engineers are often perceived as introverted intellectuals
and criticized for their use of technical terminology
that laymen have difficulty understanding. While
engineers themselves may believe they communicate
effectively with the public, my travels around the
world and discussions with the people living in various countries have led me to conclude that the public
believes otherwise. And while communicating effectively with the public is important, so is communicating effectively with members of other professions.
Because the world of engineering intersects with the
worlds of business, law, economics, finance, politics,
and most other fields within today's global marketplace, it is necessary for engineers to develop communication skills that strengthen their performance
within the complex arena of the 21st-century workplace.

Communication is simply defined in fundamental terms as the process by which information is
exchanged. Such exchange may assume many forms,

including listening, observing, reading, speaking, and writing. The most basic step in communication is listening. Unfortunately, however, our formal educational process focuses attention on reading. Little emphasis is placed on speaking and almost no attention on the skill of listening. Research has shown that two months after listening to a talk, the average listener will remember only about 25 percent of what was said. In fact, we tend to forget from one-half to one-third of it within eight hours (Nichols 1957). The importance of listening is often overlooked by engineers as a critical element of the communication process. Engineers too often make recommendations to a client before taking the time to *listen* to the client's needs or concerns. Listening is the first step in ensuring that effective communication takes place.

The purpose of communication is to convey information, and this may be done formally or informally and by a variety of means. For example, information may be conveyed via memoranda, reports, books, films, telephone conversations, electronic communications, or meetings and conferences. The art of communicating effectively entails determining the best means of conveying information to a particular audience—to an audience that is, for example, technical versus non-technical—and then selecting the form that would be most effective. The form of the communication must be carefully considered. Communication directed toward the public or toward a client, for instance, is typically more formal than, say, an internal company e-mail.

The more formal the communication, the more carefully constructed the communication must be in terms of grammar, word usage, punctuation, capitalization, and overall sentence construction—components of written communication that often are not well understood by engineers. And the more highly educated the audience, the more sophisticated the communication must be in terms of complexity of thought. Informal communication is not as demanding, although in terms of content, accuracy is of course essential.

The most common form of informal communication today is electronic mail. E-mail, however, tends to be overused or used without consideration of its potential consequences, and it can be a lazy means of communicating. E-mail also can potentially isolate individuals as well as create animosity. A case in point: Engineer A is hard at work in his office desperately trying to meet a client's deadline. Engineer B has sent to Engineer A a segment of analysis that must be incorporated into the final product, but Engineer A does not fully understand one of the conclusions in this segment. Engineer A's office is only two doors down from the office of Engineer B, but rather than walk down the hall to speak with Engineer B, he sends Engineer B an e-mail asking for an explanation. Engineer B's e-mailed response, however, only further confuses the matter. An e-mail chain continues until the two engineers end up electronically "screaming" at one another, which upsets both of them and wastes valuable time that should be spent completing the project.

E-mail has facilitated communication in many ways but sometimes a telephone conversation—or, in the case of Engineer A and Engineer B, a face-to-face conversation—is preferable. Common sense is a dimension of communication often overlooked amid the current emphasis on electronic communication. The effective communicator determines the best method of communicating given the situation and the individual(s) to whom the communication is being transmitted.

The importance of understanding the audience cannot be overemphasized. If a communicator does not sufficiently understand the audience to whom the communication is being directed—that is, if sufficient thought is not given to how a particular audience is likely to digest the information—the communication may not achieve its intended result. This understanding is absolutely key to ensuring that the audience will pay attention or "listen" to the communication. Technical jargon, acronyms, and complex words may be quite appropriate for an audience of engineers but quite inappropriate for a lay audience.

It is important to keep in mind that individuals tend to absorb only communication that is of some particular value to them. The psychologist, writer, and social commentator Hugh Mackay explored this concept in 1994 and developed ten "laws" of communication:

1. It's not what our message does to the listener, but what the listener does with our message that determines our success as communicators.
2. Listeners generally interpret messages in ways that make them feel comfortable and secure.
3. When people's attitudes are attacked head-on, they are likely to defend those attitudes and, in the process, to reinforce them.
4. People pay the most attention to messages that are relevant to their own circumstances and point of view.
5. People who are insecure in a relationship are unlikely to be good listeners.
6. People are more likely to listen to us if we also listen to them.

7. People are more likely to change in response to a combination of new experience and communication, rather than in response to communication alone.

8. People are more likely to support a change which affects them if they are consulted before a change is made.

9. The message in what is said will be interpreted in light of how, when, where, and by whom it is said.

10. Lack of self-knowledge and an unwillingness to resolve our own internal conflicts make it harder for us to communicate with other people. (Mackay 1994, 14)

So important is effective communication in today's workplace that many corporations and other entities have developed and formalized communication principles that govern all external communication. The U.S. Army Corps of Engineers, for example, has developed a communication guide founded on the following principles:

1. Listen to all constituencies, both inside and outside of the United States Army Corps of Engineers regarding issues of importance to them, respecting their viewpoint.

2. Communicate early, clearly, completely, honestly, accurately, and often with all constituencies on issues of importance.

3. Incorporate communication activities as an integral part of the project management business process.

4. Be accessible to all constituencies and respond promptly without censorship or misinformation.

5. Proactively inform the public and other constituencies of the Corps' vital role in areas in which the Corps has special expertise.

6. Do what we say we will do. (U.S. Army Corps of Engineers 2003)

Common sense is a dimension of communication that should be applied on a routine basis. It is important to give thought to the best means of communicating in any general communication situation.

Project Communication Management

Project communication principles, however, are somewhat different from general communication principles in that they are more focused on a particular project that has specific goals and objectives as well as beginning and end points. Additionally, the communication involves information that is learned on a project and conveyed to the project stakeholders. The information recorded over the course of a project is used for a variety of reasons both during the project and long after the project has been completed. Such information forms the basis of key decisions that affect project goals, scheduling, and cost. Written communication often serves to shed light on certain aspects of a project following its completion and can be used to improve the outcomes of future projects. It also may be used to support or counter claims made by various stakeholders when things do not go as planned. It is critically important for the engineer to be fully aware of the information recorded before, during, and after project execution and to ensure that the information is accurate and reflects the best data known to the engineer at the time it was recorded.

The Project Management Institute (PMI) defines project communications management as "the Knowledge Area that employs the processes required to ensure timely and appropriate generation, collection, distribution, storage, retrieval, and ultimate disposition of project information. The Project Communications Management processes provide the critical links among people and information that are necessary for successful communications" (PMI 2004, 221).

According to PMI (2004), the project communications management process consists of the following four tasks:

1. Communications planning—determining the information and communications needs of the project stakeholders
   • Who needs what information
   • When the information will be needed
   • How the information will be provided
   • Who will provide the information
2. Information distribution—providing needed information to project stakeholders in a timely manner
3. Performance reporting—collecting and distributing performance information
   • Work performance information
   • Performance measurement
   • Forecasted completion
   • Quality control measurements
   • Project management plan

- Approved change requests
- Deliverables
4. Managing stakeholders—managing communications to satisfy the requirements of and resolve issues with project stakeholders
    - Communications management plan
    - Communication methods
        - Issue logs and resolutions
        - Approved change requests
        - Approved corrective actions.

Each of these tasks is critical to successful project outcomes.

Accurate communication becomes even more critical in today's global environment. Communication among a project site, the headquarters that controls and supports the project site, and the suppliers located within other countries does not easily take place directly, and there are obstacles to effective communication—the primary obstacles being geographical distance and time differences. Thus the most common form of communication becomes the written word—typically in the form of e-mails, which have their shortcomings as we have seen.

The reliability of the communication often becomes a concern within the global marketplace as it becomes necessary to communicate not only with members of one's own firm or organization but also with those working for the members of one's consortium, owners, consultants, contractors, government authorities, and so on. The accuracy of communication with such individuals may vary widely, depending on the degree of fluency in the languages being used in the communication. This is especially true in countries where the universal business language of English is not taught widely. Engineers, therefore, must develop effective communication plans for specific projects to ensure that everyone involved is on the same page at the same time and that communications among all of those involved on the project flow smoothly.

Engineering Communication Education

Limited written and verbal skills can severely impede an engineer's performance and growth. Surveys of U.S. executives often point to effective communication skills as the single most important factor influencing promotions, and yet engineers are not perceived by and large as excelling as communicators. They tend to write and speak in a code of sorts, using acronyms, abbreviations, and technical terms that are familiar only to those with a grasp of engineering. Engineers traditionally have not had the training to communicate effectively with members of other professions or with the general public.

Engineers generally have failed to excel as communicators because of the way in which they have been educated. Those individuals who enter the engineering profession often are—or are perceived to be—analytical introverts and, thus, are not predisposed to the art of effective communication. To compound the problem, most engineering education has focused on the individual—on individual research and on individual problem solving. Teamwork and team efforts have not been emphasized. Although engineering schools across the United States have offered communication courses for more than 20 years, such courses have not been incorporated into the engineering curriculum. Engineering educators have failed to establish a correlation between communication courses—that is, composition, technical writing, and public speaking courses—and engineering coursework.

A new approach to engineering communication emerged in the United States in the 1990s. In this paradigm, engineering and communication experts work collaboratively to develop a curriculum that blends engineering and communication instruction and leverages the synergies between the two fields. This emergence of collaborative programs reflects the changes that are necessary in engineering education reform. For example, Northwestern University switched to a cross-disciplinary, cross-school collaboration in a major reform of its undergraduate engineering curriculum, requiring freshmen to take a new course entitled "Engineering Design and Communication." Instead of writing papers, the students write to faculty and clients to communicate the important information about their projects. For example, they may write mission statements, report on client meetings, synthesize the results of research, prepare progress reports, and create slides for PowerPoint presentations. Students come to understand how engineering communication combines written, oral, interpersonal, and mathematical communication.

Georgia Institute of Technology offers a course in its chemical engineering department, "Chemical Engineering 4600," that provides a new look at the importance of communication and hosts weekly guest speakers, who include state senators, graphics communicators, lobbyists, chief executive officers, patent attorneys, regulators, and others who help determine a product's success and the importance of the ability to communicate to ensure a product's success.

Standards for general education communication courses approved by the Maryland Communication Association in 1998 and endorsed by the Maryland Intersegmental Chief Academic Officers Group in April 1999 provide a useful set of general education components for consideration. The standards recommend that students taking communication courses should be able to demonstrate competency in eight areas: the communication process, verbal and nonverbal communication, message development and organization, audience and content analysis, expression, listening, analysis and evaluation, and communication ethics (Maryland Higher Education Commission 1999).

Hope College has also recognized the importance of communication. In a 2000 conference held at the college, the participants developed a report on the design of the undergraduate curriculum in communication (Hope College 2000). In addition to nine content-based curricular goals, the learning outcomes recommended as critical include understanding of multiple theoretical perspectives and diverse intellectual underpinnings in communication as reflected in its philosophy and/or history; competence in effective communication with diverse others; competence in presentation, preferably in more than one form; competence in analysis and interpretation of contemporary media; competence in reflective construction and analysis of arguments and discourse intended to influence beliefs, attitudes, values, and practices; competence in systematic inquiry (the process of asking questions and systematically attempting to answer them) and understanding the limitations of the conclusion reached; competence in analysis and practice of ethical communication; competence in human relational interaction; and competence in analysis and practice of communication that creates or results from complex social organization.

Effective communication is critical to the engineer's ability to rise above public perception and demonstrate that engineers are indeed capable of holding leadership positions. As engineering education is reformed, communication will be an essential course in preparing engineers to become effective leaders in the 21st century.

3

Ethics and Professionalism

PERHAPS NO GREATER "SOFT SKILLS" ARE NECESSARY for the engineer to acquire than the ability to deal capably with ethical issues and to behave in a professional manner, for these skills lie at the heart of the engineer's primary obligation—to hold paramount the public safety, health, and welfare. As engineers seek to enhance their image in the 21st century by achieving a better grasp of globalization and improving their ability to communicate effectively, they must also strive to enhance their image with the public whom they are obligated to protect by performing their work in accordance with ethical standards and by giving back to their profession through participation in professional activities and licensure. The engineer's role and responsibility today extends beyond protecting today's public to protecting future generations and the environment that these generations will inherit.

The term "ethics" has several meanings, but for the purposes of this discussion it refers to moral values that are sound. Despite some variances with respect to what the word "moral" may mean, a number of universal values upheld by engineers encompass what one would call to mind when speaking of ethics, and they are truth, honesty, and trustworthiness; respect for human life and human welfare, including the life and welfare of future generations; a sense of fair play; and transparency and competence. These concepts form the basis of engineering ethics and the codes of ethics under which engineers practice.

While ethics has always been an important component of engineering practice, the ethical considerations of the 21st century place a heavier burden on engineers today. Not only must today's engineers grapple with such issues as mitigating bribery and corruption—particularly but not exclusively in developing and underdeveloped nations—they must also work to devise ethical means of addressing such problems as climate change, an increase in natural disasters, and the pressing need to incorporate the principles of sustainable design into a wide array of projects. And as engineers move forward in the 21st century, they must also formulate a vision that focuses on how best to determine future societal needs and that approaches the management process accordingly, an ethical consideration not typically considered in the past.

The term "professionalism" refers to the conduct, aims, or qualities that characterize a profession or an individual professional. The late William H. Wisely, who was named the American Society of Civil Engineers' executive director emeritus in 1973 after serving the society for seventeen years, remarked that "the obligation to give primacy to the public interest is the very essence of professionalism. Without this commitment, the effort of a group to seek elite status as exponents of a body of specialized knowledge is but a shallow and selfish charade no matter how sophisticated that body of knowledge may be or how rigorously it may be pursued" (Wisely 1978).

The Engineers' Council for Professional Development states that an engineering practicing professional must possess a service motive and share advances in knowledge, safeguard professional integrity and ideals, and render gratuitous public service in addition to service rendered to clients; must recognize one's obligations to society and to other practitioners by living up to established codes of conduct; must assume relations of confidence and accept individual responsibility; and should carry one's part of professional groups as well as one's part of the responsibility of advancing professional knowledge, ideals, and practice (Mantell 1964).

Of course the root of the term "professional" is the word "profession," which may be defined variously as a calling requiring specialized knowledge and often long and intensive academic preparation; a principal calling, vocation, or employment; an occupation that requires advanced expertise, self-regulation, and concerted service to the public good; an occupation in which one is skilled; or a vocation in which professional knowledge of some level of learning is applied to serve others. But what distinguishes a profession—or more precisely, the engineering profession—from a job or an occupation? A job is a task for which one is paid, so clearly engineering is a job. An occupation is employment through which a person earns a living, and so clearly engineering is an occupation. Engineering, however, is certainly much more than a job or an occupation. While the necessity of education and training is implied in the definition of the term "profession," an individual does not become a professional simply by acquiring a broad education.

Dan Henry Pletta, who taught engineering science and mechanics at Virginia Tech for more than 40 years, noted that the purpose of the engineering profession is to serve the public by using materials and forces of nature for humanity's comfort and benefit; to provide a corps of professional practitioners eminently competent technically, whose foremost dedication is that of unselfish service to society; and to develop a nucleus of engineers to complement other professionals for the industrial, governmental, and societal leadership of our technological civilization (Pletta 1984).

The Bylaws of the American Society of Civil Engineers define a profession as "the pursuit of a learned art in the spirit of public service. . . . A profession is a calling in which special knowledge and skills are used in a distinctly intellectual plane in the service of mankind, and in which the successful expression of creative ability and application of professional knowledge are the primary rewards. There is implied the application of the highest standards of excellence in the educational fields prerequisite to the calling, in the performance of services, and in the ethical conduct of its members. Also implied is the conscious recognition of the profession's obligation to society to advance its standards and to prescribe the conduct of its members" (ASCE 2007, 21).

Charles B. Fleddermann, the associate dean for academic affairs at the University of New Mexico School of Engineering, has discussed various aspects of a profession, noting that its members perform work that requires sophisticated skills, the use of judgment, and the exercise of discretion; that membership requires extensive formal education as opposed to practical training or apprenticeship; that the public allows special societies or organizations that are controlled by members of the profession to set standards for admission to the profession and to set standards of conduct for their members and to enforce those standards; and that significant public good results from the practice of the profession (Fleddermann 2004).

Professional Engineering Organization and Licensure

A professional engineering organization serves to unify a profession and to speak on behalf of that profession on issues that affect the public. The organization also provides a forum for communicating among members, for organizing conferences and other group events of importance, and for mobilizing change within a large group. Professional engineering societies develop, preserve, and disseminate professional knowledge. Society publications and technical papers, for example, are primary sources of knowledge that enable members to keep up with advances within the profession. The meetings and conferences conducted by engineering societies afford opportunities for networking—for interaction with individuals other than the peers and executives encountered in the everyday workplace.

Professional engineering organizations also serve as powerful educational arms that effect legislation that will better protect the public and give voice to issues of significance that should be brought to the attention of policy makers. And such organizations provide guidance to educators, motivate students in the pursuit of engineering careers, and assist in the continuing professional development of engineers. In addition to the professional societies, a primary source of pride for the engineering profession is the Order of the Engineer, an association composed of graduate and professional engineers that focuses on professional ethics.

The Order of the Calling of an Engineer was established in Canada in 1926, when engineers there recognized engineers' special obligation to one another, to the profession, and to the public they serve. They decided to recognize those willing to subscribe to the code of ethics for engineers by inviting them to join the Order of the Calling of the Engineer.

Based on the Canadian model, the Order of the Engineer was established in the United States in 1970 to foster a spirit of pride and responsibility in the engineering profession, and to bridge the gap between training and experience. The first ceremony was held at Cleveland State University's Fenn College of Engineering. The ceremony involves the taking of the oath known as the Obligation of an Engineer and the bestowing of a simple steel ring—the Engineer's Ring, which is worn on the fourth finger of the working hand as a constant reminder of the engineer's primary obligation: to protect the safety, health, and welfare of the public. The history of the ring ceremony is rooted in the construction failure of a bridge across the St. Lawrence River in Quebec in 1907. Several years into the construction process, on the morning of August 29, 1907, the bridge collapsed, becoming an ugly tangle of structural steel ensnaring the bodies of 75 workmen. Following an investigation, construction recommenced but on September 11, 1916, the center span fell as it was being raised, killing 13 men. The bridge was rebuilt once again—this time without incident—and is still in use today.

Professional licensure is another means of ensuring that professional practice is maintained at the highest level; it recognizes competency and enhances the stature of engineering in the view of the public. Yet mandatory licensure has been a controversial subject for years. Licensure is mandatory in Canada, and almost all engineers in the United Kingdom and Australia are licensed, and yet—inexplicably—that is not the case in the United States. But if engineers are to protect the public safety, health, and welfare, how can they provide assurance to the public that their work is truly sound if they are not licensed?

Support of engineering licensing laws enhances the stature of the profession by holding individuals accountable for their work—both through requirements imposed by the licensing or registering body in the jurisdiction in which they work and through the legal process. Licensing should also extend into academia. If an engineering faculty member is incapable of passing a professional engineering examination that confirms the individual's competence in his or her respective teaching field, then what does that say about the quality of our engineering education? Every engineering faculty member practices engineering every day and teaches students to practice engineering. It is truly unfortunate that our profession has not considered civil engineering instruction of sufficient importance to require licensure.

Engineering faculty serve as a valuable asset to the engineering profession and serve a very important function other than that of teacher or researcher: they serve as role models to students who in turn will be the future engineering leaders of tomorrow. If engineering faculty do not take pride in the engineering profession and do not take part in professional activities and become licensed, what message does this send to students?

The stature of the engineering profession would be better enhanced if all engineers were licensed. It would be in the best interest of the profession if graduates were actively encouraged by their professors to become licensed professional engineers. A first step in accomplishing this would be to find a mechanism whereby a larger percentage of our engineering faculty was licensed. After all, the practice of engineering entails much more than the stamping of a drawing; it is, rather, an affirmation of skill and competence.

Professional Ethics

Professional ethics involves a number of considerations: the code of ethics to which one subscribes when joining a professional organization; the code of ethics to which one is bound when obtaining a professional engineering license in a specific jurisdiction; global ethics relative to how engineering is practiced; and environmental ethics regarding how the engineer begins to take into consideration sustainability issues in the design and construction of today's projects.

The concepts of ethics and professionalism are given concrete form in codes, or standards, of ethics. Codes of ethics provide a framework for defining ethical professional judgment. Such codes define the rights, duties, obligations, roles, and responsibilities of the members of the profession. However, no code of ethics is all encompassing; it is simply an aspirational document that serves as a starting point for ethical decision making.

Codes of ethics are important for two primary reasons: first, they constitute a compilation of the thoughts of members of a profession about their ethical or professional responsibilities. With respect to the engineering profession, these codes help the public understand how engineers see themselves and what engineers as a whole have learned over the years. Second, the codes are mechanisms for monitoring the behavior of engineers and for assisting them in behaving in socially responsible ways.

A code of ethics is not, however, a prescription for ethical behavior, a substitute for sound judgment, a legal document, or a tool for the creation of new moral or ethical principles. Rather, a code of ethics spells out the ways in which moral and ethical principles apply to professional practice and guides the professional in applying moral principles to situations encountered in professional practice.

The primary purpose of a code of ethics is to motivate members of a profession to conform to the principles set out in the code. Thus codes of ethics are designed to inspire, encourage, and support ethical practitioners rather than to punish wrongdoers. Codes of ethics play eight essential roles: they serve and protect the public, provide guidance, offer inspiration, establish shared standards, support responsible professionals, contribute to the education of the public, help determine wrongdoing, and strengthen a profession's image (Martin 2004).

The first professional code of ethics is believed to have been the Hippocratic Oath, which was written by Hippocrates, the father of medicine—or perhaps one of his students—in the fourth century B.C.E. The American Bar Association was the first professional organization in the United States to adopt a code of ethics, its Canons of Professional Ethics in 1908. A year later, the American Institute of Architects adopted its Code of Principles of Professional Practice and Canon of Ethics, and in 1912 the American Medical Association adopted its Principles of Medical Ethics. The first American engineering organization to adopt a code of ethics was the American Institute of Electrical Engineers in 1912. Both the American Society of Civil Engineers and the American Society of Mechanical Engineers adopted their codes in 1914. These early codes, however, contained little mention of an engineer's responsibility to the general public.

In 1936, the Japan Society of Civil Engineers established the Research Committee on Reciprocal Codes for Civil Engineers with the following objectives: to determine the duties of civil engineers, to enhance the position of civil engineers, and to promote the dignity of civil engineers. The three objectives were incorporated into a doctrine and a code of ethics that was suited to the conditions prevailing in the industry at that time, that is, May 1938. Many years later, the Asian Academies of Engineering (Chinese Academy of Engineering, the Engineering Academy of Japan, and the National Academy of Engineering of Korea) developed a code of ethics that mirrored those that had been adopted in the United States and the United Kingdom. The code of ethics adopted by the Asian engineers in November 2004 includes the following canons:

1. Accept responsibility in making engineering decisions consistent with the safety, health, and welfare of the public, and disclose all relevant information concerning public safety, health, welfare, and sustainable global development in carrying out irreversible civil engineering work [that is] long-term and large-scale in nature.
2. Act as faithful agents or trustees in business or professional matters for each employer or client, provided that such actions conform to other parts of this guideline.
3. Disclose all known or potential conflicts of interest that could influence or appear to influence their judgment or the quality of their services.
4. Be honest and trustworthy in stating claims or estimates based on available data.
5. Perform work in compliance with applicable laws, ordinances, rules and regulations, contracts, and other standards.
6. Honor [ownership] rights, including copyright and patent, and give proper credit for intellectual property.
7. Seek, accept, and offer honest professional criticism, properly credit others for their contributions, and never claim credit for work not done.
8. Give due effort to the need to achieve sustainable development and conserve and restore the productive capacity of the earth.
9. Promote mutual understanding and solidarity among Asian engineers and contribute to the amicable relationships among Asian countries. (Asian Academies of Engineering 2004)

In 1947, a major reformulation of the U.S. codes of engineering ethics was undertaken by various societies, using the Engineers' Council for Professional Development Canons of Ethics for Engineers, which mentioned a concern for the public. However, not until 1974 were these canons revised to read as they do now, stating that "engineers shall hold paramount the safety, health, and welfare of the public in the performance of their professional duties" (Baum 1980, 9).

Table 3-1 provides an overview of the various codes of ethics being used by engineering societies in the United States, the United Kingdom, Japan, and China. It is interesting to note the common ideals among these engineering societies: to hold paramount the public health, safety, and welfare; to be honest; and to encourage professional development.

TABLE 3-1. *Comparison of the Codes of Ethics of World Engineering Societies*

Value	ASCE	ABET	IEEE	ICE	FIDIC	CEAI	JSCE	AJCE	JSME
Integrity, honor, dignity	✓	✓	–	✓	✓	✓	–	✓	–
Honesty	✓	✓	✓	✓	✓	✓	✓	✓	✓
Safety, health, welfare	✓	✓	✓	–	✓	✓	✓	✓	✓
Competence	✓	✓	✓	–	✓	✓	✓	✓	–
Avoid conflicts of interest	✓	✓	✓	✓	✓	✓	✓	✓	–
Reputation	✓	–	✓	✓	✓	–	–	–	✓
Honest criticism	✓	✓	✓	✓	✓	✓	✓	–	–
Professional development	✓	–	✓	✓	✓	✓	✓	–	✓
Quality assurance	–	✓	–	✓	–	✓	–	✓	–
Disclosure	✓	–	✓	–	✓	–	✓	–	–
Confidentiality	✓	–	–	✓	✓	–	✓	✓	–
Fairness	✓	–	✓	✓	✓	–	✓	–	–
Privacy	–	–	–	✓	–	–	–	–	–
Reject bribery	–	–	✓	✓	✓	–	✓	✓	✓
Avoiding injury to others	✓	–	✓	✓	✓	–	✓	–	–
Compliance	–	–	–	–	–	–	✓	–	–
Intellectual property	–	–	–	–	–	–	–	–	–
Heritage	–	–	–	✓	✓	✓	✓	✓	✓
Assisting colleagues	✓	✓	✓	–	–	–	–	–	–
Sustainability	✓	–	–	✓	–	–	–	–	–

Note: ✓ indicates that the society's code of ethics includes the value; – indicates that it does not.

ASCE = American Society of Civil Engineers (United States); ABET = ABET, Inc., formerly the Accreditation Board for Engineering and Technology (United States); IEEE = Institute of Electrical and Electronics Engineers (United States); ICE = Institution of Civil Engineers (United Kingdom); FIDIC = International Federation of Consulting Engineers (European Union); CEAI = Consulting Engineers Association of India; JSCE = Japan Society of Civil Engineers; AJCE = Association of Japanese Consulting Engineers; JSME = Japan Society of Mechanical Engineers.

Global Ethics

Ethical standards are similar worldwide. While no two nations may define values or morals or even ethical practice in precisely the same way, some universal or nearly universal mores are applied in the global workplace: avoid exploitation, avoid paternalism, avoid bribery or the giving and/or receiving of expensive gifts, refrain from violating human rights, promote the welfare of the host country, respect the cultural norms and laws of the host country, protect the health and safety of the citizens of the host country, protect the environment of the host country, and promote a society's legitimate institutions (Harris et al. 2005).

This is not to suggest, however, that corruption and bribery do not exist, because that is most certainly not the case. But there has been a concerted effort to combat corruption and bribery on a global scale, and while certainly much still remains to be accomplished on that front, significant inroads have been made. In 1994, member firms of the World Economic Forum developed a set of global ethics for practicing in the construction industry that are known as the PACI Principles (World Economic Forum 2004). The principles were subsequently broadened to provide for multi-industry support from World Economic Forum members and nonmembers alike. The goal is widespread adoption of the PACI Principles in an effort to elevate standards across various industries and contribute to sound governance and economic development. The PACI Principles build upon general industry anti-bribery principles developed in 2002 by Transparency International—a nongovernmental organization dedicated to fighting corruption—and a coalition of private-sector interests, nongovernmental organizations, and trade unions.

In 2004, the American Society of Civil Engineers (ASCE) formed the Task Committee on Global Principles for Professional Conduct to develop principles and adopt guidelines and policies for ethical professional practice for consideration by members of engineering societies in the United States and worldwide. As part of its work, the committee reviewed the codes of more than 60 organizations worldwide and found that only two of these codes included specific clauses addressing the issue of corruption: that of the Institution of Civil Engineers in the United Kingdom and that of the American Institute of Architects. The Institution of Civil Engineers' *Code of Professional Conduct* states as a breach of rules "having any form of involvement, whether direct or indirect, and whether for the benefit of the member, the member's employer, or a third party, in bribery, fraud, deception, and corruption. Members should be especially rigorous when operating in countries where the offering and accepting of inducements and favors, or the inflation and falsification of claims, is endemic" (ICE 2004). The American Institute of Architects' *Code of Ethics and Professional Conduct* states that "members shall neither offer nor make any payment or gift to a public official with the intent of influencing the official's judgment in connection with an existing or prospective project. . . . Members serving in a public capacity shall not accept payments or gifts [that] are intended to influence their judgment" (AIA 2004).

On the basis of information supplied by the 60 organizations, ASCE's task committee ultimately drafted *Principles for Professional Conduct* for consideration by the global engineering and construction community. These principles state that

> Individual engineers can become fighters against the crime and corruption that plague the engineering and construction industry worldwide by
> - Assuring that they are not personally involved in any activity that will permit the abuse of power for private gain.
> - Recognizing that money intended for projects for the benefit of mankind is going into the pockets of dishonest individuals worldwide.
> - Understanding that corruption occurs in both public and private sectors, in both procurement and execution of projects, and among both employers and employees.
> - Refusing to condone or ignore corruption, bribery, or extortion or payments for favors wherever they are found.
> - Urging adoption of enforceable guidelines to professional practice in ethical codes of engineering societies in which they may not now exist.
> - Enforcing anticorruption guidelines by reporting infractions by members or nonmembers of the engineering profession.

Additionally, the committee proposed amendments to the ASCE Code of Ethics by enhancing the guidelines under Canon 6, which now reads:

> Engineers shall act in such a manner as to uphold and enhance the honor, integrity, and dignity of the engineering profession and shall act with zero tolerance for bribery, fraud, and corruption.

a. Engineers shall not knowingly engage in business or professional practices of a fraudulent, dishonest, or unethical nature.

b. Engineers shall be scrupulously honest in their control and spending of monies, and promote effective use of resources through open, honest, and impartial service with fidelity to the public, employers, associates, and clients.

c. Engineers shall act with zero tolerance for bribery, fraud, and corruption in any engineering or construction activities in which they are engaged.

d. Engineers should be especially vigilant to maintain appropriate ethical behavior where payments of gratuities or bribes are institutionalized practices.

e. Engineers should strive for transparency in the procurement and execution of projects. Transparency includes disclosure of names, addresses, purposes, and fees or commissions paid for all agents facilitating projects.

f. Engineers should encourage the use of certifications specifying zero tolerance for bribery, fraud, and corruption in all contracts. (ASCE 2007, 15–16)

Global Engineering Practice Ethics

Most codes of ethics devised by professional bodies attempt to identify behavior that may be unacceptable. For commercial and legal purposes, this unacceptable behavior is called "conflict of interest," a term that has been adopted by engineering associations as well. Although "conflict of interest" is used broadly in the United States and internationally, the term lacks a universally acknowledged definition, which has resulted in both real and perceived problems.

Conflicts of interest are by no means limited to construction, engineering, and related industries, but several high-profile ethical breaches have motivated domestic and international regulators to take action. New regulations could profoundly affect the way engineers do business, and if the regulations are promulgated for wide application across many different industries, engineering service companies could have difficulty complying with rules ill-suited to their business. As a result, several international organizations are taking action to address the conflict-of-interest issue. Some of their actions are described next.

International Federation of Consulting Engineers (FIDIC)

FIDIC is strongly committed to the avoidance of conflict of interest in the consulting engineering industry and to the concept that clear, transparent, and internationally accepted principles should be applied. FIDIC's policy on conflict of interest requires that consultants provide professional, objective, and impartial advice and at all times hold the client's interests paramount, without any consideration for future work and strictly avoiding conflicts with other assignments or their own corporate interests (FIDIC 2004). FIDIC Policy Statement 21 contains FIDIC's definition of conflict of interest:

> A consultant conflict of interest (COI) is a situation in which a consultant provides biased professional advice to a client in order to obtain from that client an undue benefit for himself, herself or an affiliate and in so doing, places the consultant in a position where its own interests could prevail over the interests of the client.
>
> Consultants shall not be selected for any assignment that would a) be in conflict with their prior or current obligations to other clients, or b) may place them in a position of not being able to carry out the assignment in the best interest of the client. Without limitation on the generality of this rule, consultants shall not be engaged under the circumstances set out below:
>
> a. *Conflict between consulting activities and procurement of goods, works, or services.* A firm that has been engaged by a borrower to provide goods, works, or services for a project and any of its affiliates shall be disqualified from providing consulting services related to those goods, works, or services unless the potential conflict arising from this situation has been identified and resolved in a manner acceptable to the client throughout the selection process and the execution of the contract.
>
> Conversely, a firm engaged to provide consulting services for the preparation or implementation of a project, and any of its affiliates, shall be disqualified from subsequently providing goods or works or services resulting from or directly related to the firm's earlier consulting services, unless the potential conflict arising from this situation has been identified and resolved

in a manner acceptable to the client throughout the selection process and the execution of the contract.

b. *Conflict among consulting assignments.* Consultants, including their personnel and sub-consultants or any of their affiliates, shall not be engaged for any assignment that, by its nature, may be in conflict with another assignment of the consultants unless the potential conflict arising from this situation has been identified and resolved in a manner acceptable to the client throughout the procurement and execution phases of the project.

As an example, consultants engaged to prepare engineering design for an infrastructure project shall not be engaged to prepare an independent environmental assessment for the same project, and consultants assisting a client in the privatization of public assets shall not purchase, nor advise purchasers of, such assets. Similarly, consultants hired to prepare terms of reference for an assignment shall not be engaged for the assignment in question.

c. *Relationship with the client's staff.* Consultants, including their personnel and sub-consultants, that have a business or family relationship with a member of the client's staff or of the project implementing agency's staff may not be awarded a contract, unless the conflict stemming from this relationship has been resolved in a manner acceptable to the client throughout the procurement process and the execution of the contract. (FIDIC 2004)

European Federation of Engineering Consultancy Associations (EFCA)

The European Federation of Engineering Consultancy Associations (EFCA) has been negotiating with the European Commission, the executive arm of the European Union, on the E.U. definition of conflict of interest, especially with regard to procuring consulting services. The European Commission has tended to interpret conflict of interest very strictly, while EFCA argues that the definition should be limited to several narrowly defined circumstances:

- Conflict between consulting activities and the procurement of goods or works;

- Certain conflicts within consulting assignments, for example, the preparation of terms of reference and participation in the resulting tenders;
- The execution of a project or study execution and the evaluation of the same project or study;
- The design of a project and the study of its impact on the environment;
- Advice given to both government and buyer in, for example, privatization;
- Conflict arising from family or other personal relationships. (FIDIC 2004)

In addition, EFCA argues that it is *not* a conflict of interest for a consultant who worked on one phase of a project (other than preparing the terms of reference) to be involved in a subsequent phase. Procurement will remain fair and transparent, argues EFCA, as long as all preliminary investigation documents are made available to all prospective participants.

The World Bank

The World Bank addresses the issue of conflict of interest in its Consulting Services Manual, clearly defining categories of conflict:

> A consultant conflict of interest (COI) is a situation in which consultants provide, could provide, or could be perceived as providing biased professional advice to a Borrower to obtain from that Borrower or from others an undue benefit for themselves or their affiliates. Although COI is an easily understood concept, to identify and prevent it or address its consequences (that is, the potential or actual prejudice to the Borrower's interests) requires in practice the exercise of common sense, sound judgment and expertise. Conflicts of interest must be avoided because they affect the consultant's impartiality and spoil the quality of their advice. (World Bank 2006, 16)

United Nations Educational, Scientific, and Cultural Organization (UNESCO)

In 1999, the United Nations Educational, Scientific, and Cultural Organization (UNESCO) sponsored the World Conference on

Science, which developed an ethical code of conduct for scientists. The code states that "The ethics and responsibility of science should be an integral part of the education and training of all scientists. . . . Young scientists should be appropriately encouraged to respect and adhere to the basic ethical principles and responsibilities of science" (UNESCO 2000, 481).

Association of Finnish Civil Engineers (RIL)

The Association of Finnish Civil Engineers (RIL) has undertaken a major effort focusing on the issues of corruption and bribery—with much success, apparently, as Transparency International ranks Finland as the least corrupt nation in the world. The Finnish believe that the following factors have assisted them in combating corruption and bribery: anticorruption legislation is clear and its control is open and public; limits on hospitality are commonly accepted; salaries of civil servants are in line with those of private-sector employees; public administration is transparent; and government is transferring its operative power to private enterprises to meet citizens' needs in a more customer-oriented manner (RIL, letter to ASCE President, April 13, 2005).

Environmental Ethics

Sustainable design is now requisite. Peter Kydd, a managing director of Parsons Brinckerhoff in the United Kingdom, has observed that sustainable development provides a solid foundation from which one may work to develop projects that balance environmental, societal, and economic concerns and results in an improved state for generations to come (Kydd 2006). The term "sustainable development" was introduced during the 1970s but gained currency in 1987 following publication of *Our Common Future* by the United Nations World Commission on Environment and Development. *Our Common Future* defines sustainable development as "a process of change in which the exploitation of resources, the direction of investment, the orientation of technical development, and institutional change are all in harmony and enhance both current and future potential to meet human needs and aspirations. . . . Sustainable development is development that meets the needs of the present without compromising the ability of future generations to meet their own needs" (Brundtland 1987).

In 1992, engineering organizations around the globe formed the World Engineering Partnership for Sustainable Development and formulated the following vision:

> Engineers will translate the dreams of humanity, traditional knowledge, and the concepts of science into action through the creative application of technology to achieve sustainable development. The ethics, education, and practices of the engineering profession will shape a sustainable future for all generations. To achieve this vision, the leadership of the world engineering community will join together in an integrated partnership to actively engage with all disciplines and decision makers to provide advice, leadership, and facilitation for our shared and sustainable world. (AAES 1994)

The United Nations Commission on Environment and Development has underscored the inclusion of the following fundamental components in sustainable development: environmental protection, economic growth, and social equity (UNCED 2007). John Elkington, a founder of the think tank SustainAbility in the United Kingdom, developed these three components into the "triple bottom line (TBL) concept," which requires a balanced approach to economic development, environmental protection, and social well-being—or EES for short (Elkington 1998).

The triple bottom line concept exerts its greatest impact on how engineers, planners, designers, and managers deliver a best-practice solution to their clients. Social impact will be a major consideration in all engineered projects in the 21st century. While environmental impact assessments are now common, clients will begin to require that both social–economic and human impact assessments be performed before projects can proceed. For instance, with respect to economic considerations, what is the project cost that represents the best values from the perspective of achieving the project objectives? Have the life-cycle costs been analyzed to determine the total cost of project delivery over its expected life? Have environmental factors been included in the valuation of assets and services? With respect to environmental questions, how will the project interact with the natural environment? Are there any concerns regarding the materials or products proposed that may exert potential future negative impacts on the project depending on the use or application?

When considering social impacts, the engineer should begin by asking how the person living next door is going to view the project. How can the project best be integrated into the community? Will the health, diversity, and values of the community be maintained or enhanced for the benefit of future generations? This in time will require the engineer to examine specific aspects of the project relative to its materials and products. Is the life expectancy of the selected materials and/or products the same relative to the social expectation of how long the project will function as designed? Is there a need for future inspections of any aspect of the project regarding its structural integrity to assure its sustainability over the expected design life?

Sustainability was a central component of the discussion undertaken at ASCE's Summit on the Future of Civil Engineering held in June 2006, and it was the key focus of meetings of the World Federation of Engineering Organizations a few months later. Although controversial, former Vice President Al Gore's film and book, *An Inconvenient Truth,* entrenched sustainability in the public lexicon.

The 21st century will almost certainly witness the development of new life-cycle design philosophies that could result in almost zero net waste. Almost all materials could be recycled and reused. The basis of this opinion is the fact that one of the most important political issues to gain traction during the late 20th century was environmental protection. As concern over the environmental health of planet Earth increased, the engineering profession embraced the concepts of environmentalism and sustainable practices, and as a consequence, many professional engineering organizations incorporated references to environmental protection into their codes of ethics.

Today, the principles of sustainable development are incorporated into professional engineering practice as a matter of course. However, the challenges of applying these principles on projects in developing and underdeveloped nations are still considerable. Engineers must strive to educate all elements of society and promote the universal adoption of a sustainable development ethic—particularly among developers; investors; and local, regional, and national and international governing bodies.

4

Diversity

AN ONGOING ISSUE FOR ENGINEERS IS THAT THE general public seems to have very little idea what it is that engineers actually do. In a 2004 National Science Foundation survey, only 15 percent of the respondents described themselves as very well informed about new scientific discoveries and the use of new inventions and technology. About one-third of those surveyed considered themselves poorly informed about these topics. Surveys conducted both in the United States and other countries reveal that most citizens do not have a firm grasp of basic scientific facts or concepts, nor do they have an understanding of scientific processes. No doubt, the general public would have been even less familiar with the scope of the various engineering disciplines had the surveys identified them individually.

One of the primary reasons that the public has little knowledge or understanding of engineering is that the engineering profession is perceived as remote—as far removed from the general populace. Certainly the "face" of engineering does not mirror the "face" of today's increasingly diverse populations. This problem is not an emerging issue and, in fact, has been around for quite a while. When Vannevar Bush served as director of the U.S. Office of Scientific

Research and Development, he corresponded with Presidents Franklin D. Roosevelt and Harry S. Truman and noted that both men and women should be trained in science. The engineering profession within the United States is, in the words of William A. Wulf, a former president of the National Academy of Engineering, largely a "pale male profession" (*Cornell Chronicle* 2001). While the demographics of today's general workforce have changed to reflect the diversity of the population and despite Vannevar Bush's vision in 1945, today's engineering workforce bears no resemblance to the public it serves. For example, in the 1990s, the Hispanic population in the United States increased by 60 percent and the Asian-American population increased by 63 percent. Asian-Americans are the fastest-growing segment of the population (NSF 2004). There is no clear majority of any ethnic population in California or in 50 percent of the largest cities in the United States, and minorities constitute approximately 40 percent of the populations of New York, Mississippi, Maryland, Georgia, and Arizona. Americans with a disability of some level account for 20 percent of the population (Pear 2005). By 2050, Asians, Pacific Islanders, blacks, Hispanics, and American Indians/Alaskan natives are expected to account for more than half of the resident college-age population in the United States (NSF 2004). As the pace of globalization increases, the impoverishment of our "pale male" profession becomes more pronounced.

What would be required in the United States to change the engineer's image from "pale male" to "diverse"? The answer to this question is surprisingly simple: the engineering profession must become much more diverse in its composition. To this end, engineering must be explained much as other professions are explained within the U.S. educational system, beginning at the kindergarten level and continuing through high school. An engineering career should be portrayed—and explained—much as the medical, legal, architectural, and accounting professions. Students must come to understand at an early age what engineers do, why their work is so important to our society, and how exciting an engineering career can be—and they must come to understand that they, too, can become engineers. The onus is as much on engineers as it is on teachers to ensure that students realize this. Engineers must visit their local schools and assist teachers in explaining what engineering is and in describing what engineers do, so that the diverse students in U.S. schools understand why engineering is interesting and fun and how engineering helps the public and the planet, improving the quality of life for all.

In October 1998, Congress established the Commission on Advancement of Women and Minorities in Science, Engineering, and Technological Development (CAWMSET). The mandate was to research and recommend ways to improve the recruitment, retention, and representation of women, underrepresented minorities, and persons with disabilities in science, engineering, and technology education and employment. As was noted in the preamble to the commission's final report,

> Today's U.S. economy depends more than ever on the talents of skilled, high-tech workers. To sustain America's preeminence we must take drastic steps to change the way we develop our workforce. An increasingly large proportion of the workforce consists of women, underrepresented minorities, and persons with disabilities—groups not well represented in science, engineering, and technology (SET) fields. Unless the SET labor market becomes more representative of the general U.S. workforce, the nation may likely face severe shortages in SET workers, such as those already seen in many computer-related occupations. (CAWMSET 2000, iii)

The report goes on in the Executive Summary to say,

> If, on the other hand, the United States continues failing to prepare citizens from all population groups for participation in the new, technology-driven economy, our nation will risk losing its economic and intellectual preeminence. It is time to move beyond a mere description of the problem toward implementation of a national agenda that will take us where we must go, so that our nation can thrive now, and in the years to come. It is also time to establish clear lines of responsibility and to define effective accountability mechanisms. (CAWMSET 2000, 1)

The term of the commission has expired, and now—seven years after this report was issued—little action has been taken on these recommendations.

In a 2007 report, a committee of the National Academies noted that the vitality of the United States economy is derived in part from the productivity of well-trained people and the steady stream of scientific and technical innovations they produce. But the report warns that the erosion of U.S. leadership in the fields of science and technology is threatening the nation's preeminent position within the global economy. "We have led the world for decades," the report notes, "and we continue to do so in many research fields today. But the world is changing rapidly, and our advantages are no longer unique" (Committee on Prospering 2007, 12–13). The report's recommendations focus on the need to improve science and mathematics education at the kindergarten through high school levels, increase research, improve higher education, and broaden economic policy to provide incentives for innovation.

Such efforts constitute critical steps in strengthening the nation's capabilities and in heightening the public's awareness and understanding of engineering. But the need to effect change in the ways in which the United States *develops* its workforce is essential and how this need is addressed will affect America's future. As Joseph Bordogna, the former deputy director and chief operating officer of the National Science Foundation, noted in an address to the Engineering Workforce Commission:

> We need a robust mix. . . . Demographic factors and questions of capitalizing on all available talent tell us that a larger portion of the engineering workforce must be women, underrepresented minorities, and persons with disabilities. Said another way, we need to attract more of our diverse domestic talent to the engineering fold. . . . The differences that abound in race and ethnicity and gender in our society should be gainfully embraced. . . . The largely untapped pool of women, underrepresented minorities and persons with disabilities is our nation's competitive "ace-in-the-hole." While doing better than we did thirty years ago, we have not yet seriously capitalized on this national treasure and are now playing catch up in a very competitive world. (Bordogna 2003, 2)

While the demographics of the American workforce have changed to reflect the diversity of the overall population, the composition of the American engineering workforce of today bears no resemblance to the public it serves. In the 1990s, for example, the Hispanic population of the United States increased by nearly 60 percent. Most of this growth was experienced in four states—California, Texas, Florida, and New York—although the Hispanic population extends throughout the United States. Similarly, the Asian-American population of the United States grew by 63 percent during the 1990s and is the most rapidly growing segment of the American population. According to the National Academy of Engineering, half of the U.S. population will be nonwhite by 2050, and engineers will increasingly serve diverse communities (NAE 2005).

The National Science Board's *Science and Engineering Indicators 2006* notes that although Asians–Pacific Islanders constitute only five percent of the U.S population, they accounted for seven percent of college graduates (NSB 2006, 3–20). The *Science and Engineering Indicators 2006* noted that the percentage of historically underrepresented groups in science and engineering occupations remained lower than the percentage of those groups in the total college-educated workforce. Women made up 24.7 percent of the science and engineering workforce and 48.6 percent of the college-educated workforce. Blacks made up 6.9 percent of the science and engineering workforce and 7.4 percent of the college-educated workforce. Hispanics made up 3.2 percent of the science and engineering workforce and 4.3 percent of the college-educated workforce (NSB 2006, 3–19).

How must the engineering profession go about changing the face of engineering to reflect the face of the population? The steps are straightforward. The engineering profession must improve its communication with the public in terms of explaining what engineering is, what engineers do—and in terms of explaining what an engaging profession engineering is. Engineers continue to define themselves as problem solvers, but they are also in equal measure creators and innovators who are addressing the great challenges of our time. Engineering is, in essence, about improving the quality of life for the public, and engineers must put forth a concerted effort to communicate how this is being done by the various engineering disciplines. Initially, engineers will require assistance from professionals in the media and communication fields to get this message across because engineers themselves are not yet equipped to effectively communicate this message. Such efforts will require considerable assistance from professional engineering societies, which must speak for the profession as a whole. This assistance presents challenges of its own, however, in that professional societies have yet to mount adequate public relations or media campaigns to this end—either because they lack the professional staff adequately skilled to do so or because they lack the funding to do so.

In addition to seeking advice from professionals regarding how best to present a sound engineering message to the public, engineers must immediately become actively involved in working with educators at the kindergarten through high school levels to inspire and challenge their students to pursue careers in engineering. Real engineers speaking to students about real engineering experiences goes a long way in leaving lasting positive impressions about engineering as a career choice. Engineers must also determine why more students—young girls in particular—are not attracted to engineering as a career even though they have chosen a technology-based career path. And engineers must understand the importance of mentoring young engineers—particularly young women—to better the retention rate of these engineers.

A 2004 poll of American adults conducted by Harris Interactive found that most would be extremely pleased if their children pursued a career in engineering, yet just one-third of those same adults considered themselves to be well informed about engineering. In fact, this poll revealed that the engineering profession generally falls in the middle of the prestige rankings. Engineering ranked tenth among 22 occupations in the survey, with 29 percent of the public rating its prestige as "good"—about the same level as in a 2003 Harris poll—but was down from 34 percent in 2002 and 36 percent in 2001 (NSB 2006, 7–38).

The Gender Gap

Particularly troublesome is the gender gap within the engineering profession in the United States. Engineering in America is indeed a "pale male" profession. Catalyst, an organization that conducts studies of women in a variety fields, notes several reasons why women choose to leave the engineering profession: many female graduate students within the fields of science remain uninformed about potential careers in business, academia is viewed by many as unwelcoming to women scientists, and there is an absence of female role models. Other reasons include isolation, risk-averse supervisors and stereotypes, differences in style, exclusion from informal networks, lack of mentoring, and work–life balance issues (Catalyst 1999).

In addition, according to a similar 2003 Catalyst study, factors that hold women back from reaching the top include failure of senior leadership to assume accountability for women's advancement; exclusion from informal networks; lack of significant general management experience; stereotyping and preconceptions of women's roles and abilities; and commitment to personal or family responsibilities (Catalyst 2003).

In 2004, Catalyst published another report, which established a quantitative link between women's representation in senior leadership and corporate financial performance. The 353 companies studied that remained in the Fortune 500 for four of five years between 1996 and 2000 boasted a higher representation of women in senior management positions than the companies that dropped off the list. The report also notes that the group of companies with the highest representation of women on their top management teams experienced a return on equity that is 35.1 percent higher than the companies with the lowest responsibilities. The same group saw a total return to shareholders that is higher by 34.0 percent (Catalyst 2004). However, the engineering profession has yet to learn from these proven facts and leverage the full potential of the nation's diversity. A decade later, little has changed relative to why women leave the engineering profession.

The engineering profession—and society as a whole—suffers from this gender imbalance. Engineers are viewed as quintessential problem solvers. Give an engineer some parameters, a few constraints, and a target to aim for, and away he or she goes, not to see the light of day again, so to speak, until the problem is "solved." However, "solved" has different meanings, depending on who is devising the solution. Engineers tend to see a solution solely from their individual perspectives. Thus, if that perspective does not include a sufficiently diverse base upon which the problem is analyzed, the solution will be lacking key elements.

A case in point is the design of the 1999 Ford Windstar minivan, sold by the Ford Motor Company. For years, women who chose to stay at home and raise their children complained that minivans did not adequately meet their needs or the needs of their children. Ford Motor Company theorized that the reason for this was that the minivans were designed by men, not by the women who drove them on a very frequent basis. Ford's solution was to assemble a team of engineers composed of both men and women and led by a woman. When Ford created this development team in 1997, it had 30 female engineers (known as the Windstar Moms) at a time when only 8.5 percent of all engineers were women (U.S. Bureau of the Census 1995–2000). The women on the team had owned minivans and understood what sorts of features mothers needed in a minivan. The result was the Windstar, an enormously popular minivan that featured numerous safety features and was the first such vehicle to incorporate "sleeping baby lights"—lights that do not shine from overhead into a sleeping child's eyes (About.com 2007; Maynard 1999).

Attracting Young Girls to the Profession

Many young people today—especially young girls—require role models and mentors to provide them with the hands-on guidance and encouragement that will prompt them to consider a career in engineering.

In a 2003 report, *New Formulas for America's Workforce: Girls in Science and Engineering,* the National Science Foundation suggests that many girls need more than basic classroom exercises to become fully engaged in a course of study or career that, given its imposing edifice of mathematics, may appear unduly abstract. Hands-on learning is a proven tool for improving learning on the part of all students and is indispensable in instilling self-confidence in girls and in stimulating their interest in science and engineering. Supplemental programs that combine hands-on activities with exposure to female role models are essential in attracting young women to engineering and sustaining their interest in it.

To strengthen its scientific, engineering, and technical workforce, the United States must reach out to young women, who on the whole constitute a very small portion of it. Women constitute 46 percent of the United States workforce, but only 23 percent of the science and engineering workforce (Jackson 2003). This paucity means that valuable perspectives and experiences are missing in America's pursuit of advancements in science and engineering.

Why are girls not enrolling in college engineering degree programs and going on to pursue engineering careers in larger numbers? According to a U.S. Department of Education study (Huang et al. 2000), the problem is not one of ability. Despite long-held perceptions, many girls express much interest in math and science and perform as well as or better than their male counterparts in these subjects. The report found that girls are taking high school science and math courses in approximately the same numbers as boys: 94 percent of girls and 91 percent of boys take biology; 64 percent of girls and 57 percent of boys take chemistry, 26 percent of girls and 32 percent of boys take physics, and 64 percent of girls and 60 percent of boys take algebra II.

Yet few girls choose to pursue an education and subsequent career in engineering. There is much speculation as to why girls are not choosing engineering. Various explanations for this have been proffered, among them that high school girls are not aware of many women working in engineering and thus few role models exist, are not aware of what a career in engineering entails and view engineering as a career for introverted intellectuals, perceive engineering to be a male profession, and do not view engineering as an enjoyable profession that would provide them with the opportunity to earn a good salary and enjoy personal flexibility.

Research conducted by the Japan Society of Civil Engineers in 2004 revealed that girls are not interested in engineering because they associate it with difficult, dangerous, and dirty work environments and long hours and that they believe they would be in the minority and thus would have little in common with their coworkers. They also believe that they would not be treated equally and would have restricted work hours and work locations and would be subjected to other differential treatment (JSCE 2004).

And yet in 2002, for the first time in history, the top elected leaders of four of the five oldest U.S. engineering societies—the American Society of Civil Engineers, the National Society of Professional Engineers, the American Society of Mechanical Engineers, and the Institute of Electrical and Electronics Engineers—were women, a feat unmatched by even the legal and medical professions. While this unique accomplishment clearly demonstrates that women have shattered the glass ceiling of professional achievement within the engineering profession, it also struck a startling counterpoint—albeit a positive one—to statistics confirming the low number of women entering the profession.

These female leaders and their male counterparts recognized the importance of diversity to the future of the engineering profession, and through their participation in summits coordinated by the American Association of Engineering Societies and hosted by the National Academy of Engineering, their respective organizations joined with several other engineering organizations to sign pledges supporting diversity within the engineering profession. As a result, a committee on diversity was established within the American Association of Engineering Societies to serve as the coordinating body for engineering societies and other related organizations to conduct collaborative programs focused on increasing representation by women and other underrepresented groups.

The benefits of such collaboration to advance engineering outreach were formally recognized in a 2002 report published by the National Academy of Engineering (NAE 2002). The report was the outcome of a study commissioned in 2001 that would assist the engineering community to maximize its resources in delivering a "comprehensive, coordinated, and sustained message" (NAE 2002). The report cited as one the most effective engineering outreach programs National Engineers Week, an effort coordinated by a coalition of engineering societies with support from major engineering corporations. Among the report's chief recommendations was that the engineering community focus its outreach efforts on the precollege audience.

In response to this recommendation and to counter the misperceptions of high school girls and the people who influence them, WGBH Boston—a public television station responsible for such television programs as *Nova*, *Building Big*, and *Zoom into Engineering*— the American Association of Engineering Societies, and a coalition of more than 80 engineering and related associations formed a partnership to develop and implement the Extraordinary Women in Engineering Project. Consisting of a national outreach campaign, a television broadcast, a companion Web site, and a large-format book, the initiative was designed to mobilize the eight million engineers throughout the United States to actively reach out to educators and girls and encourage participation in engineering education and careers; to provide compelling role models of women actively engaged in fulfilling engineering careers; to help high school science, math, and technology teachers and guidance and career counselors better understand the nature of engineering, the academic background needed by students pursuing engineering, and the career paths offered by engineering; and to equip high school teachers and counselors to share this information with students, especially with girls.

The Extraordinary Women in Engineering Project conducted a study of more than 5,000 high school girls from a diverse population across the United States. Interviews of these girls revealed a profound lack of interest in engineering as well as little understanding of the engineering profession. These girls perceived the engineering profession as a male-dominated profession that would afford them little opportunity, as evidenced by the very small number of role models. Most alarming were comments indicating that the girls wanted to enter a profession "that helped people," which indicated they had no understanding that "helping people" is what engineering is all about.

Through hands-on activities and exposure to role models, the Extraordinary Women in Engineering Project initiative worked to spark an interest among young girls in engineering, science, and technology while also helping them gain the confidence necessary to consider careers in these areas. The starting point was the 2006 book

Changing Our World: True Stories of Women Engineers. The book offers profiles of outstanding women engineers—past and present—who have significantly contributed to a better world (Hatch 2006). By inspiring and retaining young women in this way and others, the engineering profession will secure the next generation of role models for girls seeking a challenging, rewarding career.

The Next Steps

Engineers must engage in some radical thinking about education and about their public image. They must focus on how engineers will be required to think and work in a world in which complexity and change are not occasional events and in which the workplace is continuously responding to new technologies and to the 24/7 schedule made possible by the Internet. Engineers must rethink how they educate prospective engineers, beginning at the kindergarten level and proceeding through the university level and beyond.

Women members of the Japan Society of Civil Engineers believe that the steps to be taken to increase the number of women entering the engineering profession include improving the work environment, altering the male consciousness to become more cognizant of professional equality, educating men on the importance of sharing in child-care responsibilities, promoting women engineering role models, countering existing stereotypes within society, creating awareness of and encouraging discussion of gender difference issues, and continuing to gather statistics on the number of women entering and remaining in the engineering profession (JSCE 2004).

The 21st-century engineer must resolve to encourage individuals from all diverse groups to enter into the complex and dynamic field of engineering lest the profession squander the opportunity to maximize the potential for an infusion of intellectual capital. As Joseph Bordogna said in a 2004 speech:

> It is *not* about the total number of scientists and engineers the nation may or may not need. It's easy to get distracted by debates about trends and statistics that attempt to make the case that the demand for science, engineering and technology workers is greater or less than the supply. Rather, what it *is* about is drawing into the engineering and science workforce a larger propor-

tion of women, underrepresented minorities and persons with disabilities, no matter the workforce size. *Whatever the numbers turn out to be, we need a robust and varied mix*, and that means broadening participation. (Bordogna 2004, 5)

5

Leadership

LEADERSHIP IS MOST SIMPLY DEFINED AS THE CAPAC-
ity to lead. However, leadership is also the ability to
influence others for the purpose of positive gain—
with or without the designated authority to do so.
The traits generally viewed as conducive to leadership include commitment, confidence, communication, curiosity, entrepreneurship, a strong work
ethic, an insistence on excellence, honesty, integrity,
sound judgment, persistence, a positive attitude, and
sensitivity. Desirable behaviors of leaders that can
be taught and learned include earning trust, communicating well, thinking rationally, being open and
consistent, demonstrating a commitment to organizational values, and using discretion when handling
sensitive information.

Edward J. Hoffman, the director of the National
Aeronautics and Space Administration's Academy of
Program/Project and Engineering Leadership, empha-
sizes four central components of leadership:

1. *The desire to win:* successful leaders will
endow their missions with a passion, enthusiasm, and commitment to succeed that is
contagious.

2. *A focus on results:* effective leaders make certain that their teams understand the requirements and objectives that lead to project success.

3. *The ability to adapt to change:* good leaders establish a climate in which team members can respond productively to uncertainty or a shift in objectives.

4. *The ability to create an environment of trust:* successful projects rely on collaboration, which requires all team members—including employees, vendors, consultants, and partners—to trust one another. (Hoffman 2005, 22)

Leadership also entails transcending the ordinary to become someone whom others aspire to emulate. Leadership is about capturing the imagination and enthusiasm of others and bringing out the best in them both individually and collectively. At one time, U.S. engineers held significant leadership positions and were highly respected by the general public. George Washington, Theodore Roosevelt, Herbert Hoover, and Jimmy Carter, for example, were engineers who became U.S. presidents. These engineers who went on to assume the highest elected office in the United States consistently demonstrated to the public their commitment to the betterment of society while never losing sight of what engineering contributed to their development as leaders. In one of his often-cited speeches, President Herbert Hoover described the responsibilities and rewards of the engineering profession thus:

> It is a great profession. There is a fascination of watching a figment of the imagination emerge through the aid of science to a plan on paper. Then it moves to realization in stone or metal or energy. Then it brings jobs and homes to men. Then it elevates the standards of living and adds to the comforts of life. This is the engineer's high privilege. The great liability of the engineer compared to other professions is that his works are out in the open where all can see them. His acts, step by step, are in hard substance. He cannot bury his mistakes in the grave like doctors. He cannot argue them into thin air or blame the judge like lawyers. He cannot, like the architects, cover his failures with trees and vines. He cannot, like the politician, screen his shortcomings by blaming his opponents and hope that the people will forget. The engineer simply cannot deny that he did it. If his works do not work, he is dammed. (Hoover 1961, 131–134)

The public perception today, however, is that engineers are not capable of being leaders—that they simply do not possess the requisite skills. Unfortunately, that perception is accurate, primarily because many engineers do lack the "soft" skills required for effective leadership. While engineering practice today requires formal training and experience, it also requires engineers to acquire skills that will reestablish them as leaders in the public's eye. Today's engineers simply must understand that if they do not acquire the "soft" skills needed to assume leadership positions they will not succeed in the 21st-century global marketplace.

What is the profile of today's successful leaders? Such leaders assume command of a situation and develop a course of action that ensures measurable progress and ultimate success. They possess the skills to develop strategy, think laterally, and lead by example while empowering a team or teams with the responsibility to carry out and complete tasks satisfactorily and while motivating team members through personal enthusiasm and charisma. Good leaders also learn from their mistakes. The Order of the Engineer, discussed in chapter three, has its origins in a tragic engineering mistake. In 1907, the Quebec Bridge over the St. Lawrence River in Canada collapsed, killing 75 workers and injuring 11 others. The collapse was attributed to engineers' failure to perform additional calculations to confirm the structural integrity of the span. All engineers learned and continue to learn from that mistake.

In 1940, the original Tacoma Narrows Bridge—situated on the Tacoma Narrows in Puget Sound near Tacoma, Washington—collapsed as a result of wind-induced vibrations. This collapse brought engineers worldwide to the realization that aerodynamic phenomena in suspension bridges were not adequately understood by the engineering profession and had not been adequately addressed in the bridge's design. Those who investigated the collapse recommended the use of wind-tunnel tests to aid in the design of the second Tacoma Narrows Bridge, and these recommendations resulted in the testing of all existing and future suspension bridges in the United States.

An error of far greater magnitude led to the 1981 Kansas City Hyatt Regency walkway collapse. One of the distinguishing features of the Kansas City Hyatt Regency was a multistory atrium that was traversed by suspended concrete walkways at the second, third, and fourth levels—the fourth-level walkway situated directly above the second-level walkway. The engineer had accepted the contractor's change on the shop drawing for the walkway structural support rods because the revision would reduce installation cost and construction time. Tragically, this revision, which doubled the load on the connection between the fourth-floor walkway support beams and the tie rods carrying the weight of the second-floor walkway, resulted in increased static loads on the connections one July evening, when approximately 2,000 people were gathered in the atrium—dozens of them on the walkways—to observe a dance contest. When the live load exceeded the capacity of the structural support rods, the connection failed, and both walkways collapsed onto the atrium floor, killing 114 people and injuring 200 more.

Engineers who overlook or sidestep the importance of shop drawing review in the field or who are not mindful that a change to a shop drawing during construction can result in a fatal error like the Hyatt Regency tragedy must assume responsibility for their flawed judgment. Engineering leaders recognize the critical importance of design accuracy and precision and understand that there is absolutely no excuse for taking shortcuts in engineering practice.

Leadership versus Management

Leaders do not "manage" people, they inspire and guide them. Leaders are visionaries and strategists. Leaders maintain an open mind and listen to others. Leaders develop their own individual approaches to motivating and persuading others. Although the work engineers perform is important, the manner in which engineering leaders conduct themselves significantly affects the way their efforts are understood and appreciated by others. Leaders do not assume and do not merely rely on information they are told. Instead, leaders gather as much information as possible, separate the facts from fiction, and recognize how to understand the details without micromanaging. Leaders decide what must be done as well as how it must be done, decide the path an organization should follow, and set goals for achieving an organization's objectives. Leaders then motivate the organization's managers to execute established goals and objectives. And they do so by determining the strengths and knowledge of individual team members and then maximizing these strengths and this knowledge. In *The Engineering Profession*, Dan Henry Pletta presents his formula for successful leadership:

1. Convey a sense of mission.
2. Keep superiors informed.
3. Prepare a worst-case scenario.
4. Keep things simple and take note of early warnings.
5. Seek help quickly when it is needed.
6. Take reasonable risks, and do not become obsessed with failure.
7. Cultivate a how-to-cope attitude.
8. Do not lose touch with coworkers, and listen to their opinions.
9. When criticism becomes necessary, administer it in small doses at first and do not humiliate individuals in front of others.
10. Motivate subordinates by assigning challenging opportunities and praising their achievements.
11. Be fair; strict, but fair.
12. Strive for excellence.
13. Make decisions and accept responsibility.
14. Issue directions that facilitate acceptance without resentment.
15. Recruit and train capable subordinates to prevent group failures. (Pletta 1984, 196–197)

The primary distinction between leadership and management is that leadership entails the development and execution of a vision whereas management entails the oversight of individuals and other resources to effect the execution of that vision. Managers may in due course become leaders, but that path is not assured; leadership requires considerable creativity while management does not. But leadership does require an understanding of some principles that are applicable to the oversight of individuals—principles used to lay the foundation of effective, productive organizations and teams, which play critical roles in the execution of a vision. These principles fall within four categories: effective communication, value principles, synergy principles, and investment principles (Cottrell 2002).

The components of *effective communication* for a leader are building trust; honoring commitments, or keeping promises; sharing knowledge; and providing feedback. A leader must gain trust from team members; failure to do so will compromise the work of the team, because if there is a lack of trust, whatever the leader communicates to team members will be viewed as suspect. A leader must also honor commitments—for example, must consistently meet deadlines—or the team will lose confidence in the leader's ability. Leaders must also share their knowledge with team members so that team members learn from the leader and understand the leader's vision for any given undertaking. Feedback to team members is critical. Not only do individual team members learn whether or not they are meeting expectations, they also learn how their performance is or is not contributing to the overall team goals and objectives.

Value principles form the foundation of earning respect and are categorized as principles of integrity, principles of responsibility, principles of commitment, and principles of vision. Principles of integrity ensure that what is accomplished by the leader and by the team is accomplished honestly. Principles of responsibility ensure that the leader accepts the consequences when things go wrong and does not blame other team members. Principles of commitment guide the team in achieving excellence; the team leader makes a commitment to a job well done and, by example, inspires all of the team members to meet a standard of excellence. Principles of vision provide each and every team member with a clear view of the leader's vision so that each member can embrace this vision.

Synergy principles constitute the glue that holds the team together and include the principle of communication, the principle of conflict resolution, the principle of optimism, and the principle of change management. Effective communication is essential to team cohesion. When team members have a clear understanding of the leader's expectations—the result of effective communication on the part of the team leader—they are more likely to perform well. And when conflict arises between team members, a strong leader knows how to address it. Conflict is not necessarily problematic. Opposing points of view can result in the advancement of new ideas or new solutions to problems. It is important, however, for the leader to know when and how to negotiate with individuals who cannot resolve their conflict so that the team's goals and overall performance are not jeopardized. Optimism is infectious and goes a long way toward bolstering team confidence and morale and sustaining enthusiasm for the team's objectives. Strong change management skills are critical to effective leadership in the engineering profession because engineering projects seldom unfold precisely as planned.

Investment principles are the intangibles that enhance teamwork and include the principle of empowerment, the principle of courage, the principle of good example, and the principle of preparation.

Empowerment is simply providing a team member with an opportunity and the resources to use it. It conveys the sense that the leader has confidence in an individual's ability to achieve. For the purposes of this discussion, courage is defined as not fearing the difficult situations that arise during the course of every project and as standing up for one's beliefs. It also means not backing down during encounters with difficult individuals in powerful or leadership positions. The principle of good example means that it is the leader's role to lead by example. For instance, the leader does not leave until the day's work is complete. The leader shoulders as much work as the other team members and, more often than not, usually shoulders more work than the other team members. The result is that the team works more efficiently and more productively because the members are giving their all to the team effort. Preparation is synonymous with doing one's homework. The key to success is preparation and hard work, and the effective leader prepares thoroughly prior to meeting with a client, for example. Being well prepared very often determines whether one secures a contract with a client.

The Art of Mastering Leadership Skills

Almost all engineering education is technical; engineers are not taught how to lead, they are taught how to become engineers. Engineers must be taught to be leaders by supervisors who mentor them, or they must acquire leadership skills on their own initiative. While the engineering leader's skill set embraces a number of competencies, there are three basic keys to leadership success that can be easily applied to the engineering profession: communication, confidence, and commitment.

Communication

Communication is discussed extensively in chapter two relative to the basic communication skills necessary for engineers and earlier in this chapter relative to the supervision of individuals. The communication skills required by engineering leaders are discussed here in a more specific sense with regard to how an engineer communicates effectively with fellow engineers within a professional organization; with other employees within a professional organization; with students in the classroom; and with the public, lawmaking bodies, and clients.

Prior to delivering a speech or making a presentation, for example, the engineering leader must do his or her homework to gain an understanding of the audience's or client's background and specific needs. People have more respect for those who take the time to understand specific situations and specific needs. Speaking and writing should be undertaken with both appropriateness and sensitivity. Bringing clarity to complex material is a key component of the engineering leader's communication. While a more formal approach may be appropriate when one addresses lawmakers, the public, and/or the media, a more informal approach may be more appropriate when one speaks to employees, because such an approach serves to demonstrate collegiality.

Engineering leaders must also recognize the fact that engineering work is not performed in a vacuum, but rather is accomplished through the efforts of a team. Engineering leaders must understand that the team's accomplishments are dependent upon the leader's ability to motivate team members individually and collectively. Leaders must therefore share information so that each team member understands the goals and objectives of the team's undertakings. A leader should never assume that team members are fully aware of his or her expectations. Expectations must be clearly communicated to each team member. It is also important to bear in mind that accessibility, collegiality, civility, and empathy on the part of a leader go a long way toward establishing and maintaining good team/employee morale.

Good ideas are generated by individuals, but great ideas are accomplished through team discussion and deliberation. Communication plays a key role in how teams best accomplish their tasks. In today's global economy the engineering leader further recognizes that teams will be diverse and that the strength of the team is dependent upon this diversity. The good engineering leader also recognizes that in the end the credit for the outcome will be awarded to several minds, not just one. True leaders are not those who strive to be first but those who are first to strive and who give their all for the success of the team. True leaders are the first to determine a need, envision a plan to satisfy that need, and empower a team to carry out that plan. The power of the team is unleashed by the strength of the leader's commitment. As Alexander Graham Bell once noted, "Great discoveries and improvements invariably involve the cooperation of many minds. I may have been given credit for having blazed the trail, but when I look at the subsequent developments, I feel the credit is due to others rather than myself."

Quotes such as Bell's are useful motivational tools. As engineers may not yet be the best of communicators, it is often helpful to know that one can use the words of great leaders or achievers to inspire others. To cite some examples:

- Walt Disney: "It's kind of fun to do the impossible."
- Eleanor Roosevelt: "The future belongs to those who believe in the beauty of their dreams."
- Charles Kettering: "The Wright Brothers flew right through the smoke of impossibility."
- Hubert Humphrey: "Never give up and never give in."
- Walter Bagehot: "A great pleasure in life is doing what people say you cannot do."
- André Gide: "He who makes great demands on himself is naturally inclined to make great demands on others."
- Bruce Jenner: "I learned that the only way you are going to get anywhere in life is to work hard at it. If you do, you'll win; if you don't, you won't."
- Thomas Edison: "Many of life's failures are people who did not realize how close they were to success when they gave up."
- John Wayne: "Courage is being scared to death but saddling up anyway."
- Bruce Barton: "Nothing splendid has ever been achieved except by those who dare to believe that something inside of them was superior to circumstance."

Confidence

Leadership, of course, requires more than effective communication. It requires confidence—the ability to know that you can do anything if you put your mind to it—as well as the ability to never accept "no" for an answer, but rather look upon the naysayers and the apparent impossibilities as challenges and opportunities. A leader must possess the confidence in himself or herself that will ensure that work will be performed successfully and that team goals will be accomplished. It is imperative that any individuals assuming engineering leadership positions fully understand the need to convey a level of confidence in all that they do to everyone around them.

Confidence requires courage—the courage to set high personal and corporate standards, to maintain equilibrium in the face of major setbacks, to hold individuals accountable for their responsibilities in their performance of assigned tasks, to confront those who exhibit unacceptable behavior and to provide them with guidance for correcting such behavior, to resist the temptation to take work for the sake of taking work, to uphold one's reputation and integrity and the reputation and integrity of one's team and one's firm, and to take risks with the understanding that great rewards may result but that so too may failure, and that failure is a great learning device.

Commitment

Too many engineers come close to making new discoveries only to be thwarted by their frustration at the duration of the discovery process, giving up before their discovery is made. Success is realized largely by overcoming frustration, delays, setbacks, obstacles, or disappointments—in other words it is realized by means of commitment. A leader establishes goals, develops action plans for accomplishing those goals, and commits to the follow-through needed to accomplish the goals. Goal-setting is the foundation of commitment. Both personal and corporate goals should be written down. Goals should be near term and long term—typically one-year, five-year and 10-year goals. Goals should be periodically reviewed to determine progress and whether strategy modifications are necessary to reach those goals. Goal-setting is a lifelong exercise and an activity that should be communicated to others. Failure to follow through on goals can lead to a lack of confidence in oneself. In the event that others find this follow-through lacking, they too will begin to lose confidence in the leader. A loss in confidence is typically followed by a loss in trust, and once lost, trust is difficult to restore.

It is committed engineers who become successful leaders. If the engineering profession is to change the public's perception that engineers are not capable of being leaders, then engineers must change themselves. Engineering leadership does not have to begin at a national level; it can be initiated at the individual level, the team level, and the firm level. As Margaret Mead once observed, "Never doubt that a small group of committed people can change the world. It is the only thing that ever has."

Evaluating Your Own Leadership Skills

The checklist that follows was prepared to help individuals conduct an introspective assessment of their leadership skills by asking themselves a series of simple questions (Maxwell 2003).

1. *How much influence do I have?* A leader must exert influence on team members, meaning that a leader must possess qualities that serve to motivate others. A job title does not define an individual as a leader; the individual defines himself or herself as a leader. People look to leaders for guidance, and this holds true both up and down the hierarchy of a team or a staff.

2. *How solid are my "people" skills?* In other words, how well do I work with others? Do others like to work with me? This is not to be confused with the question, am I liked? There is a distinction between the terms "like" and "respect." Leaders are not necessarily liked but they are respected, and respect is something that is earned, not bestowed. Business relationships are not about friendship, they are about collaboration and consistent productive interaction. Leaders inquire about the "personal" side of their team or staff and remember what's important to their employees as individuals and not just as workers.

3. *Do I maintain a positive attitude?* Am I someone who sees the glass as half full rather than half empty? People are more likely to follow optimists than pessimists. Do I focus on possibilities rather than on obstacles? Do I have a can-do attitude? Do I focus on solutions rather than on problems?

4. *Do I possess self-discipline?* Do I organize my work, plan my day, complete assignments on schedule? Do I address problems as they arise? Do I insist on accuracy? Is my work thorough? Am I willing to work as many hours as necessary to complete a project on schedule? Am I willing to place the good of the team ahead of my own immediate desires?

5. *Do I succeed more often than I fail?* What have I accomplished? What have I learned from my mistakes? What have I learned from the mistakes of others? What would I like to accomplish this year? Within the next five years? The next ten years? What do I hope to learn from this job? What do I hope to learn from this organization? Am I making consistent, mea-

surable progress in my career? Am I being rewarded for my efforts by my superiors? Am I consistently assigned increasingly challenging projects? Am I consistently setting new goals for myself? Do I take risks? Do I seek opportunities in new challenges?

6. *Do I solve problems effectively?* Am I capable of dissecting problems to determine the root or roots of their cause? Do I seek advice from others when I cannot solve a problem on my own?

Strategies for Leadership Success

Two of the greatest leaders of the 20th century—General George S. Patton, a leading military officer during World War II, and Sir Ernest Shackleton, a celebrated Antarctic explorer—shared in common several strategies for strong, effective leadership that can be distilled as follows (Perkins et al. 2000):

1. Never lose sight of the ultimate goal.
2. Set an example with memorable symbols and behaviors so that people want to emulate you.
3. Instill optimism but stay grounded in reality.
4. Take care of yourself; maintain your stamina and let go of guilt.
5. Continuously reinforce the team message.
6. Minimize status differences and insist on courtesy and mutual respect.
7. Master conflict, deal with anger in small doses, and avoid needless power struggles.
8. Find something to celebrate and to laugh about.
9. Be willing to take big risks.
10. Never give up.

Six additional strategies go a long way toward establishing effective leadership (Axelrod 1999):

1. Always do more than is required. Individuals who demonstrate that they will do more than they are asked to do will be valued.

2. Clock-watchers and those who do not regularly work beyond the nine-to-five workday seldom rise above middle management.
3. Success is an attitude. Those who believe they will win usually do, whereas those who believe they will fail probably will.
4. Leadership is often a matter of balancing time against available resources. Ideal conditions seldom exist.
5. Choose to be surrounded by individuals who possess knowledge and sound judgment and are trustworthy. Such individuals are typically achievers.
6. Offer praise publicly and criticism privately.

Ultimately, leadership is achieved by embracing an open-minded and flexible approach, supporting team members, providing training and mentoring to team members, communicating the vision and goals for an undertaking, and acknowledging and celebrating achievements and successes. Successful leadership is the result of developing a vision, inspiring people to produce their best work, building and motivating high-performance teams, breaking down individual and organizational barriers, stimulating and promoting innovation, and cultivating enthusiasm. As Colin Powell, a former United States Secretary of State and decorated military general, once observed, "There are no secrets to success; it is the result of preparation, hard work, and learning from failure." A true leader has the confidence to stand alone, the courage to make difficult decisions, and the empathy to consider the needs of others. He or she does not set out to be a leader, but rather becomes one by the quality of his or her actions and the integrity of his or her intent.

Leaders are much like eagles: they are appreciated individually rather than in flocks.

6

The Engineer's Role in Public Policy

ENGINEERS HAVE HAD LITTLE SAY IN THE FORMULA-
tion of strategies driving some of the most important
initiatives introduced within the past decade—those
aimed at maintaining a livable world. Instead, the for-
mulation of these strategies has fallen to public policy
experts, economists, attorneys, and environmental-
ists. The issues involved are of enormous import and
include climate change, the maintenance of infrastructure, the availability of potable water, waste management, the response to natural disasters, and other significant health and safety concerns. Despite the fact
that science and technology are the tools necessary to
adequately address these issues, engineers—who are
masters of science and technology—have not been at
the forefront of the efforts to do so.

To a large extent, engineers themselves are to
blame for their lack of influence. As individuals or
members of national professional organizations,
they simply have not become active participants in
undertakings that are calling attention to the need for
reforms in how the world is built and how resources
are utilized and conserved. Engineers have ceded those
leadership roles to others, allowing them to develop
strategies for reform and determine the routes that

nations will take while engineers content themselves with building the locomotives and laying the rails, so to speak.

In 2003, the American Association of Engineering Societies commissioned Harris Interactive to conduct a poll that would provide a better understanding of the U.S. public's attitudes toward engineering. The results of the poll indicate that doctors and scientists received the highest rankings out of 22 occupations (Harris Interactive 2004). They were the only occupations that were seen by more than half of the respondents as having great prestige. Firemen and teachers tied for third, followed by military officers, nurses, police officers, and priests. Engineering fell within the middle rankings, with only 29 percent of the public saying it held the profession in high regard. The trend among the public is not good, however, given that engineering was ranked as high-prestige by 34 percent of the respondents in a different 2002 survey and by 36 percent in 2001 (NSB 2006, 7–38).

The work of engineers is not well known to the public partially because of engineers' reluctance to become involved in the public policy process. For example, only a handful of engineers currently serve in the U.S. Congress. Engineers have failed to recognize the direct link between the public policy process and the engineer's role to protect the safety, health, and welfare of the public. They mistakenly believe that as nonprofit organizations, professional engineering societies cannot lobby or speak for the profession to members of Congress and that engineers or members of professional engineering organizations cannot hold office or assist in political campaigns. They fail to understand the importance of moving their professional organizations into the 21st century by stepping up lobbying efforts and scheduling meetings with lawmakers. They have also seemed reluctant to speak out on public policy issues. But as Norman R. Augustine, a retired chief executive officer of the Lockheed Martin Corporation and a former member of the President's Committee of Advisors on Science and Technology, remarked, engineering is a "stealth" profession and if engineers do not speak out on their behalf, who will? (Augustine 1994). He told engineers at a 1998 meeting of the American Institute of Aeronautics and Astronautics,

> The time has arrived when engineers will have to venture out from the shelter and comfort of the Ivory Tower and enter the arena of boiling controversy, real world debate, and—brace yourselves—politics. It is no longer viable to place our high-tech

candle under a bushel, for at best we will find ourselves in dark-
ness and at worst our bushel will go up in flames. . . .

Engineers must become as adept in dealing with societal and
political forces as they are with gravitational and electromagnetic
forces. We must equip engineers of the future to present their
cases in almost every forum imaginable—from town meeting to
state legislature, from *The New York Times* to *60 Minutes*, from
the Congress to the Oval Office. If, as in the past, engineers place
their trust solely in the primacy of logic and technical skills, they
will lose the contest for the public's attention—and in the end,
both the public and the technical communities will be the losers.
(Augustine 1998)

Engineers are in fact ideally suited to the public policy process.
The word "engineer" is derived from the Latin verb *ingeniare*, which
means to design or create. Engineers are creative people who invent,
develop, and manage the manufacture of their creations, which range
from computers to robots to space vehicles. Engineers are characterized by their continuous quest for knowledge, and their work revolves
around this quest. Engineers are the people who have advanced humanity into the era of spaceflight and Internet information technology. The
engineering mind-set is to define a problem, identify solutions to the
problem, select the best solution, and then implement this solution.

Engineers as Politicians

Engineers' commitment to protect the public safety, health, and welfare—combined with their ability to devise solutions to complex problems—makes them ideally suited to public service and political office.
Additionally, engineers typically possess a superior grasp of current
scientific issues that career politicians typically lack, and this knowledge can serve them well in debating legislation governing such issues
as clean water, energy policies, and air pollution. There are in fact
many notable examples of engineer public servants—among them the
U.S. presidents George Washington, Theodore Roosevelt, Herbert
Hoover, and Jimmy Carter.

A most interesting case study of the success engineers can achieve
in public office can be found not in the United States but in the Dmitrov
District of Russia, which is located approximately 70 kilometers from

Moscow. The district consists of two towns, four settlements, and 440 villages and encompasses 1,000 kilometers of roads. The head of the district—an elected office comparable to that of a U.S. governor—is a professional civil engineer by the name of Valeri V. Gavrilov, who has held three five-year terms. (It is perhaps worth noting that the late Boris Yeltsin, first president of the Russian Federation, was also a civil engineer.) In 2004, Gavrilov told a delegation of representatives from the American Society of Civil Engineers (which included myself), "Happy is the man who has the profession of a civil engineer—who can leave behind what he has created with his own hands and can tell his grandchildren that he built this. Happy is the man who has the profession of the civil engineer who will leave his mark on the world for the people to enjoy after he has long past gone."

Gavrilov is a man of great vision and has been eager to see services in his district expand to improve citizens' lives. Upon the collapse of the Soviet Union, he made certain that the Dmitrov District was one of the first areas in the Moscow region to undergo development. It was understood early on that private investment would be required to improve conditions there and to foster a business climate. Russia may have gone through difficult times in the 1990s, but that was not the case in Dmitrov. Industrial plants remained open, and jobs were plentiful. More than 4,000 workers were needed in the region as a result of rapid expansion. Unemployment stood at one percent, and that figure represents the portion of the population that, for whatever reasons, simply does not care to work. Young people are encouraged to stay, and facilities have been constructed for family activities and athletic events. A new sports complex was built, and 20 more are planned for the district. The quest to engage young people in activities has significantly reduced juvenile delinquency and drug use and created a base for the city's workforce.

The bureaucracy that held sway during the Soviet regime severely restricted how civil engineering projects were undertaken and built. The goal was simply to satisfy basic needs. Cost was the principal factor, and quality was given short shrift. Outside assistance was inconceivable. Upon the fall of the Soviet regime, massive reconstruction was therefore called for, and multiple problems had to be addressed. Particular attention had to be given to housing. During the Soviet era, housing was of course owned and controlled by the state. Projects to house the population were carried out relatively swiftly then, but the speed came at the expense of quality. And following construction, there was minimal investment in maintenance. Those buildings must now be demolished and are being replaced with apartments, duplexes, and single-family homes designed in such a way as to situate parks, libraries, stores, sports complexes, and day care centers within walking distance.

Private investment and a capital market have created a new middle class whose members own automobiles and earn salaries sufficient to provide them with disposable income. Gavrilov has long recognized that partnerships with companies and outside sources provide greater scope for innovation and make it possible to expand the infrastructure.

Innovation has been a key ingredient in the changes seen in Dmitrov. Fountains, for example, have been constructed to provide entertainment, the water "dancing" to music—both traditional Russian music and popular music from around the world. Partnerships have been forged with nations worldwide to foster technological exchange. Agriculture has become an important business, and joint ventures have been formed with a number of countries to carry out research in this area. Moscow, with its population of 10.5 million, is nearby, so there is a ready market for crops and animal products. The only research station in Russia focusing on farm breeding is to be built in Dmitrov. San Diego, a sister city, will be providing assistance in the construction of a major medical center. Venture capital from the United States has been used to construct a nursing home and a senior center, their construction of very high quality.

Roads and bridges are being repaired and brought up to international standards. Additional roads are being constructed. Buildings of historical importance—including a number of cathedrals—are being restored to their original state, and structures from the Soviet era are being demolished to make way for apartment buildings, houses, and industrial facilities. The design and construction in progress involve firms from around the world.

Making the Transition

Public policy, globalization, and professionalism are all components of Gavrilov's vision—a civil engineer's dream. Engineers worldwide should strive to embrace this dream and bring to fruition their own version of it. How difficult would that be? Can engineers actually make a successful transition into the political arena? The answer is yes, and the reason lies in how engineers think. Engineers develop solutions to problems by analyzing the problems, devising various solutions to the problems, and then arriving at the best solution to the problems. Politicians, on the other hand, analyze problems and select the solution to the problem that will satisfy their constituents' desires or needs. Thus engineers clearly hold the advantage: while politicians may make decisions that involve compromise, engineers do not. Engineers make decisions that provide the optimal solutions to problems. Engineers, therefore, would be likely to ensure that the welfare of the public is not compromised, while ensuring that government decisions are made in the best interest of the nation. Taking this rationale a step further, government involvement is not only essential to the engineer's responsibility, it is essential to the survival of the engineering profession. Government is vital in upholding the standards of the profession and enhancing its integrity. Government wields the power and influence to advance projects from the drawing board to reality. If the engineer were to take a major role in the regulatory and legislative processes, not only would the engineering profession benefit but so too would the public the profession serves.

If engineers are to raise the bar for their profession, they must view their personal involvement in the public policy process as a career enhancement. Engineers must develop a greater awareness of what their professional engineering organizations can accomplish in the public policy arena as well as an awareness of what they as individuals can do. While some professional organizations cannot endorse specific candidates for office because it would jeopardize their tax status, most can and do and actively participate in public policy initiatives and lobbying with respect to issues that impact the engineering profession. As an individual, an engineer can run for office, participate in political campaigns, and make contributions to those running for office whom the engineer believes would best serve the nation and the engineering profession.

On a global scale, the engineering profession must dispel the notion that engineers cannot participate in public policy or politics simply because they are engineers. Engineers sell themselves short with their reasoning that they would not be qualified for political appointments or are not sufficiently knowledgeable about the issues to write letters to their representatives in Congress. Engineers are in fact more qualified than most to do both. During a speech delivered at Purdue

University in 1995, John H. Sununu, a mechanical engineer, former governor of New Hampshire, and former chief of staff to former President George H.W. Bush, said,

> People don't think of engineers as being actively involved in public policy, but I think there is a great need for people with the quantitative skills and the problem-solving understanding that engineers have to be much more visible, active and involved. Public policy is getting more and more dependent on technology. . . . Nobody ever expects a mechanical engineer to end up as governor or chief of staff to the president of the United States. I think my background served me well in government, and I try to encourage some of the young engineers who might be listening to be a little bit more interested in taking that route. I understand there is only one real engineer in the U.S. Senate, and [he] happens to be my son. (Sununu 2003)

Richard G. Weingardt, a civil engineer, author, and former president of the American Council of Engineering Companies, wrote that

> engineers cannot afford to sit on the sidelines while others shape our physical environment and public policy. By virtue of our training and experience, we're well qualified to apply innovative problem-solving skills in the public arena. Getting involved in government enables us to take the lead in addressing critical quality-of-life issues facing American communities: crumbling infrastructure, environmental and economic decline, public transportation, hazardous waste, and crime. (Weingardt 1995)

Even if they choose not to run for political office, engineers can communicate with the lawmakers in their respective countries. The Institute of Electrical and Electronics Engineers, Inc., has developed the following guidelines to assist engineers in communicating with their lawmakers:

- Identify clearly the issue in which you are interested. Be sure to include the House or Senate bill number if addressing specific legislative proposals.
- State briefly why you're concerned about the issue. Your personal experience will lend supporting evidence. Explain how

you think the issue will affect your business, profession, community, or family.

- Explain how your issue or concern affects the Congressman's constituents and how you think those constituents will benefit from your position.

- If you want your Congressman to take action on your behalf, clearly (but politely) ask for this action. Do not expect Members of Congress to know exactly how to solve the problem, and do not expect them to read between the lines to discern what you want done.

- If you have an idea you would like to see turned into legislation, suggest this initiative to your Congressman. Volunteer your services as an information resource or researcher on the subject.

- If your issue has been discussed in newspapers or magazines, be sure to include copies with your correspondence. If the issue hasn't been included in the news media, it might be useful to attract the interest of the press first.

- Restrict yourself to one topic in a letter or other form of communication. Concentrate your arguments, summarize them, and make your recommendations on one page.

- Use your own words and avoid technical terms. Also avoid using trite phrases or clichés, which can make your correspondence sound mass-produced. When Members of Congress receive many letters with nearly identical wording, they may discount them as being part of an organized pressure campaign. This method works only when mail is so voluminous that it has to be weighed. Personalized, individual letters often work best.

- Communicate at any time, but especially when legislation is being considered by Congressional committees or subcommittees, before it reaches the House or Senate floor. Your communication will mean more when attention is currently focused on the subject matter of your concern.

- Found out the committees and subcommittees on which your Congressman or Senators serve. They have more influence over legislation in these jurisdictions.

- Present the best arguments in favor of your position and ask for their consideration. You may find it useful to review arguments against your position and show why your position is preferred over others.

- Communicate with Members of Congress as a constituent, not as a self-appointed neighborhood, community, or industry spokesperson. However, if you are truly representing a particular group, mention it.
- If Senators or Representatives have supported your cause or idea in previous legislation, let them know that you appreciate their past leadership on the issue and that such support is applauded by their constituents. (IEEE 2005)

But while communicating with lawmakers and policy makers is important, becoming a lawmaker or policy maker is even more important. Policy makers and the public would benefit enormously from an understanding and appreciation of engineering, and thus engineers have an obligation to participate in public policy. In an effort to maximize engineers' effectiveness in the public policy arena, engineering societies should work collaboratively and leverage their resources. Engineering societies should be the advocates of the engineering profession's common viewpoints on issues important to their respective nations. In this way engineering societies can contribute effectively in shaping public policy.

If engineers continue to shy away from the public policy process, they will place the engineering profession in jeopardy. Engineers must take a much more active role in shaping public policy to ensure that legislation is enacted that is truly in the interest of protecting the public safety, health, and welfare. It is of the utmost importance that 21st-century engineers assume leadership positions in the public policy arena and serve as powerful, positive influences in the shaping of public policy and in the administration of government.

7

Engineering
Education Reform

IN THE 21ST CENTURY, AN EVER-INCREASING NEED
will emerge for a holistic breed of engineer—one who
can work across borders, cultural boundaries, and
social contexts and who can work effectively with
nonengineers. As the trend toward a more global
and more knowledge-based society continues, the
practice of engineering must be changed, and this
change must be accomplished through engineering
education reform. The engineering curriculum can
no longer remain as it has for essentially the past 40
years. The subjects of globalization, diversity, world
cultures and languages, communication, leadership,
and ethics must constitute a core component of the
overall engineering education just as physics and
mathematics do.

For nearly 20 years, numerous journals, articles,
reports and studies have been prepared by the American Society for Engineering Education, National
Academies Press, the National Science Board, the
National Science Foundation, and the American Society of Civil Engineers (ASCE 2001) that discuss the
critical need for change in engineering education. Yet
despite this large library of literature on the subject
of engineering education reform, the engineering curriculum of today still does not provide the foundation

necessary to ensure the engineer's success in the 21st century. A report by the American Electronics Association in 2007 notes that the 21st-century global economy portends to be very different from that of the 20th century and that national public policy must be viewed through the prism of education as a lifelong process (AEA 2007). In the past, the skills workers acquired would serve them well for decades. In the 21st century, however, the success of engineers and firms will be measured against how well they can adapt to new conditions and technologies. Thus to remain competitive in this global and knowledge-based economy and to ensure that the quality of life improves for everyone around the world, engineers must be educated differently.

The content of engineering education was diminished across the globe over the course of the past century in an effort to reduce the cost burden to students and to produce more engineers and to process more revenue through the educational business faster. While this devolution resulted in an emphasis on the technical aspects of engineering, it eliminated other critical aspects of engineering education and studies that are fundamental to the very essence of engineering, which is to improve the quality of life and to protect the public safety, health, and welfare. If an engineer is not trained in public policy, ethics, leadership, communication, and management, an engineer cannot adequately serve the public.

In *Educating the Engineer of 2020*, the National Academy of Engineering notes that

> today, the practice of engineering needs to change further because of the demands for technologies and products that exceed existing knowledge bases and because of the changed professional environment in which engineers need to operate. That change must be encouraged and facilitated by change in engineering education. (NAE 2005, 13)

In an article published in the magazine of the National Society of Professional Engineers, Ralph R. Peterson, the chairman and chief executive officer of CH2M Hill, a global engineering firm, wrote:

> It's not enough to simply design and build projects. We must 1) grasp the totality of our client's mission, 2) develop solutions that add value to the client's mission, and 3) link our compensation to the value-added outcomes as defined by our clients. . . . No

single firm can move the industry, but if our profession takes on the challenge . . . together we can make a difference. . . . As the developing and developed worlds strive to improve their economic prospects and quality of life, we will come to grips with the fact that we live in a finite world. Our profession's leadership challenge is to reposition itself as an effective steward of natural resources and the environment. (Peterson 2005).

The evolving global environment in which the 21st-century engineer must succeed demands that engineering education be reformed. While the basics of engineering education will not change—that is, the mathematical and scientific fundamentals—the skill set of engineering graduates must be much broader than it is today. Engineers must be much better equipped not only to function in the global economy but to flourish in it, and their education must include instruction in communication, in multiteam participation, in the design of complex systems, in multiculturalism, and in languages.

Engineering education reform will require engineers themselves to demand such reform—engineers working in academia and engineers working in industry. In other parts of the world, engineers are celebrated as heroes and leaders, and in such countries as China and India engineering enrollment is increasing significantly. In the United States, in contrast, engineers are viewed as technology enthusiasts slavishly devoted to intellectual or academic pursuits. As AEA notes, this outlook is extremely unfortunate in that it is rooted in ignorance, and ignorance is poisonous to an economy fueled by technology and innovation. Engineers must determine a way to foster a public image of innovation and creativity that is so inspirational that the young people of today yearn to pursue careers in engineering. Our educators must instill within their students the belief that engineers are engaged in creative, stimulating, challenging, and satisfying work that significantly improves the lives of people the world over. If the engineering profession is to succeed in attracting, preparing, and retaining greater numbers of qualified individuals to the profession, it must reform its educational system by shattering stereotypical images and strengthening and extending engineering education so that it is comparable to the demands of a legal or medical education, for example; by broadening the content of engineering education so that engineering graduates are confident in their ability to work effectively on multicultural teams and ultimately supervise multicultural and multidisciplinary teams;

and by including instruction in management, leadership, ethics, and professionalism so that they can better serve the public in whichever nations they find themselves working.

In 2004, Judith Ramaley eloquently made the point that engineering education today must foster "richer social networks" in an effort to broaden engineering education to instill a deeper understanding of civil responsibility and international justice as well as a deeper understanding of human capacity. She recommended an approach to engineering education that

- Develops the intellect and the capacity and inclination for life-long learning;
- Shapes ethical judgment and the capacity for insight and concern for others, the world in which we live, and the future we will bequeath to our children and their children;
- Fosters an increased understanding and openness to other cultures, languages, and societies and the connections that bind us together as fellow travelers on an increasingly connected globe;
- Builds an understanding of the effect of the human presence on the land and the effects of the systems we build to advance our efforts as a community;
- Expands our scientific horizon and our appreciation of the influence of new technologies on our lives, both as individuals and as we live them in community with others;
- Supports our capacity to nurture democratic and global knowledge and engagement, and . . . hardest of all . . . helps us acquire the ability to reach out to our adversaries and those who seek to harm us and to understand why they act as they do. (Ramaley 2004)

Inadequacy of the Four-Year Bachelor's Degree

A comparison of the historical educational requirements for the legal and medical professions serves as an ideal starting point for formulating a platform for engineering education reform. In 1900, the U.S. medical profession required of its graduates three years of college academics followed by a one-year internship. At that same time, the legal profession required two years of collegiate study. The engineering profession—in stark contrast—required *four* years of college study and was considered the most demanding profession in which to gain entry. But not only did the engineering profession fail to maintain a competitive pace with the legal and medical professions during the ensuing century, it actually reduced the number of credit hours required to earn an engineering degree. A few decades ago, the average number of credit hours required for an engineering degree was 140; today—in more than half of American colleges and universities—the number of credit hours required is 128. The engineering profession is no longer comparable to the legal or medical professions in terms of educational demands, and this is one reason that its image on the professional stage has been diminished. The public shares this view: engineering is not perceived as being as "tough" as law or medicine in terms of educational rigor.

Engineers themselves have grown complacent in terms of their education, failing to grasp the importance of a well-rounded educational experience. As Brent Strong observed, there is a disturbing tendency among engineers to simply comply with general educational requirements and not seek a real understanding of the world—an interdisciplinary understanding based in liberal arts and humanities that provides enrichment. Strong suggests that engineers strive to become Renaissance men and women (Strong 2003).

To some extent this is what the medical profession did by requiring medical students to understand human behavior and to develop a sensitive bedside manner so that they could more effectively treat their patients and establish a sense of trust. Additionally, the medical profession determined that general practice would require four years of preprofessional collegiate study followed by four years of professional study and one year of internship. Physicians aspiring to practice as specialists would be obligated to undertake an additional two to three years of study followed by residency in a hospital. The legal profession also increased its educational requirements; aspiring attorneys must complete four years of collegiate study followed by three years of study in law school.

The objectives of engineering education simply cannot be achieved within a four-year period of study. As Dan Henry Pletta observed in *The Engineering Profession*, the objectives to which the engineering profession should aspire should be

- To educate novices for the responsible practice of a specified professional art;
- To transmit applicable existing knowledge after first "distilling" it for concise presentation;
- To search for new knowledge that enhances the art involved;
- To convey a sense of ethics and professionalism;
- To motivate novices for public advocacy roles to protect the public health, safety, and welfare as well as the earth's resources and its environment; and
- To groom societal leaders for a technological civilization that will protect freedom. (Pletta 1984, 138)

The diminishment of engineering education requirements is perhaps rooted in the splintering of the profession into areas of specialization that include, for example, civil, mechanical, chemical, nuclear, mining, electrical, and metallurgical engineering. In 1980, ABET, Inc., listed 21 programs in engineering and 48 in technology. Unlike the legal and medical professions, which speak via one primary professional organization respectively, engineering speaks via dozens of specialty organizations. The engineering profession must compensate for this splintering by making a concerted effort to reestablish unity and focus on shaping educational curricula that educate engineers who can succeed in the 21st-century global marketplace and knowledge-based society.

Today's engineers are faced with challenges that are vastly different from the challenges faced by previous generations. These challenges include global commercial competition, intelligent technology, and a constantly changing work environment. These demands require knowledge that cannot be acquired by means of a four-year curriculum; they require knowledge acquired via graduate study. It is critical for the engineering profession to understand that the four-year curriculum is no longer adequate—that given the rate of change within the field of technology and the need to cope with the increased breadth and complexity of modern engineering practice, additional subjects of study must be incorporated into engineering curricula at both the undergraduate and graduate levels. These subjects include management and the subsets of finance, leadership, and human resources; probability; teamwork; communication—both oral and written; interdisciplinary project work; and foreign language and cultural study.

Heretofore, engineering education has emphasized technical content and has all but ignored professional obligations to the public.

Engineering education has largely been developed by educators rather than practitioners, but collaboration between the two segments is essential if engineering curricula are going to impart a thorough understanding of what is required for an engineer to best serve society today. The consistent lack of emphasis on professionalism, ethics, and leadership—even within doctoral programs—has hampered engineers in their service to society.

In *Engineering the Future of Civil Engineering* (ASCE 2001), the American Society of Civil Engineers notes significant and rapid changes confronting the profession. Almost all discussion of educating the 21st-century engineer presumes additions to the engineering curriculum that would include courses on communication, social sciences, business and economics, cross-cultural dynamics, and information technology. Unfortunately, the typical undergraduate engineering program requires 10 percent more course work than nontechnical degree programs and thus adding these new elements to the current four-year bachelor's program is not a viable solution. *Engineering the Future of Civil Engineers* recommends that the civil engineering profession revamp its 150-year-old educational system to meet the challenges of the 21st century. Specifically, it recommends education beyond the four-year bachelor's degree that will provide the knowledge, skills, and attitudes necessary to ensure high professional standards and protect the public safety, health, and welfare.

In *Educating the Engineer of 2020: Adapting Engineering Education to the New Century*, the National Academy of Engineering recommends that the four-year engineering bachelor's program be considered merely a pre-engineering or engineer-in-training degree and that a master's degree be considered the professional degree (NAE 2005).

Engineering Accreditation in the United States

During the 1920s, the American Association of Engineers established the Committee on Accredited Schools and called for more comprehensive and discriminating standards for evaluating engineering schools. Recognition of the need to ensure the quality of education prompted the National Council of Examiners for Engineering and Surveying and a number of engineering organizations to form a joint committee to establish improved standards for accredited schools. That committee called for the formation of the Engineers' Council for Professional Development, which later became known as the Accreditation Board for Engineering Education and Technology, or ABET. ABET is an organization of member societies in engineering and engineering-related fields. As part of its mission, ABET accredits educational programs and promotes quality and innovation in education.

In 1995, ABET published a draft of *Engineering Criteria 2000* as a mandate to educators to design curricula that could produce engineers with adequate skills with which to enter the job market. On November 2, 1996, the ABET Board of Directors approved what was initially known as the *Engineering Criteria 2000* but is now known simply as the ABET engineering criteria. These criteria include a set of 11 outcomes that all engineering baccalaureate graduates should achieve. The authors of these criteria concentrated on what all engineers should be able to accomplish regardless of their discipline of practice. The criteria can be divided into two categories: hard skills and soft skills. The six soft skills have come to be known as the "professional skills" and include an ability to function on multidisciplinary teams; an understanding of professional and ethical responsibility; an ability to communicate effectively; the broad education necessary to understand the impact of engineering solutions within a global, economic, environmental, and societal context; a recognition of the need for and an ability to engage in lifelong learning; and a knowledge of contemporary issues.

Although the curricula of many colleges and universities have been tweaked—or, in some cases, overhauled—in response to ABET's criteria, there is some frustration that this has not happened quickly enough. The most difficult piece to address has been the ability to communicate effectively. Historically, those who have been attracted to engineering have not been deemed to be naturally proficient in terms of communication. The reality of today, however, is that an understanding of the technical aspects of engineering will get an engineer only so far. Engineers must now demonstrate proficiency in technical writing, oral presentation, social interaction, consensus building, and survival on the political landscape of business.

According to the American Society of Engineering Education, employers seek engineers who demonstrate an ability to apply their knowledge of mathematics, science, and engineering to design and conduct experiments and analyze data; an ability to perform on multidisciplinary teams and communicate effective solutions within a global and societal context—which demands a grasp of everything from history to sociology to psychology; a yen for lifelong learning; and a bona fide knowledge of contemporary issues. In an issue of the National Academy of Engineering's *The Bridge* that was devoted to reforming engineering education, Latucca et al. reported on a survey of 1,622 employers. Barely half of these employers considered engineers' understanding of organizational, cultural, and environmental contexts and constraints to be adequate. Moreover, skills in these areas, according to employers, appear to have declined within the past decade. Graduates' teamwork and communication skills were rated just "adequate" (Latucca et al. 2006).

In another article in the same issue of *The Bridge*, Theodore C. Kennedy writes,

> We have to change what we expect from engineers, and we have to turn out graduates with broader skills, interests, and abilities. With the commoditizing of basic design engineering and the migration of that function overseas, the traditional training ground for recent graduates is no longer available in the United States. . . .
>
> I look for different skills than I did 10 years ago. Today, it is not unusual for good candidates to have global references and experience on projects and assignments around the world. I think we must prepare our graduates for that type of career. . . .
>
> I need graduates who can speak before an audience to make a point, either to me or a client. Comfortable or not, engineers today are constantly selling—selling an idea, a concept, a study, an alternative, or just the need for a new document control system. . . .
>
> Engineers must be able to write reports, studies, or routine business letters . . . I am tired of "cite," "sight," and "site" being used interchangeably. (Kennedy 2006, 15–16)

The ABET criteria constitute a solid first step toward reforming engineering education. It is not possible, however, to provide to engineers within a four-year period all of the skills they need to practice. The master's degree is essential to professional practice today. It is essential for ABET—and other accreditation institutions the world over—to recognize this fact and promote the benefits of the master's degree. This is the only way in which the engineering profession can protect its members from becoming commodities in the world market and gain the public trust and professional recognition that will elevate the engineering profession to a higher plateau.

The Next Steps Forward

In 1957, the Soviet Union's *Sputnik* satellite was launched, sparking an aggressive crusade to improve math and science education in the United States. In *The World Is Flat: A Brief History of the Twenty-First Century*, Thomas L. Friedman observes that the generation of scientists and engineers who were motivated to pursue careers in science by the threat that *Sputnik* represented are reaching their retirement years and are not being replaced in the numbers needed to sufficiently advance the United States within the fields of science and technology so that it remains competitive with other nations (Friedman 2005). In 2007, fifty years after *Sputnik's* launch, the United States is once again facing the threat of scientific diminishment—at a time when engineering will nonetheless continue to grow increasingly complex. As a result of the rapid rise of information technology, the explosion of knowledge within the field of engineering, and the growing complexity of civil infrastructure systems, the job performed by the engineer will continue to become more demanding. Thus, engineers today must possess both greater breadth of capability and greater specialized technical and managerial competence. Engineering practice today requires engineers to take a broader view of their work environment and to interact with the public and with policy makers on a regular basis. This interaction requires effective communication skills as well as strong negotiation skills.

Twenty-first-century engineering practice demands fundamental change in engineering education. Most of the senior members of the profession are graduates of baccalaureate programs that required the successful completion of between 145 and 160 credit hours for graduation. The norm today ranges from 120 to 135 credit hours, and these requirements continue to be reduced steadily by various universities and legislatures. How can engineers continue to do more with essentially less education? They cannot, and if engineers within academe and industry do not work collaboratively with government to effect change, engineering will cease to be a profession and will become a trade on the world commodity market. The practice of civil engineering has become increasingly more complex technically within the past 30 years—and will continue to increase in complexity—and yet the technical content of the undergraduate curriculum is not keeping pace with this complexity.

The American Society of Civil Engineers took a bold step in 2004, when it released its report *Civil Engineering Body of Knowledge for the 21st Century: Preparing the Civil Engineer for the Future.* The current version of the body of knowledge assumes the form of 15 outcomes, which represent much more technical and professional practice capability than is found in today's civil engineering programs. Of those fifteen, the technical desiderata are as follows:

- An ability to apply knowledge of mathematics, science, and engineering;
- An ability to design and conduct experiments, as well as to analyze and interpret data;
- An ability to design a system, component, or process to meet desired needs;
- An ability to identify, formulate, and solve engineering problems;
- An ability to understand the techniques, skills, and modern engineering tools necessary for engineering practice;
- An ability to apply knowledge in a specialized area related to civil engineering;
- An understanding of the elements of project management, construction, and asset management.

With respect to professional practice, the outcomes would be as follows:

- An ability to function on multidisciplinary teams;
- An understanding of professional and ethical responsibility;
- An ability to communicate effectively;
- The broad education necessary to understand the impact of engineering solutions in a global and societal context;
- A recognition of the need for, and an ability to engage in, life-long learning;
- A knowledge of contemporary issues;
- An understanding of business and public policy and administration fundamentals;
- An understanding of the role of a leader and of leadership principles and attitudes. (ASCE 2004, 5–6)

The second edition of the report is due in 2008 and includes additional outcomes. This body of knowledge—or "raise the bar"—initiative is intended to apply to all engineering graduates seeking licensure, not just to those who choose to or are able to attend graduate school. It is anticipated that in the future those earning bachelor's degrees will be able to attain the additional required education in a variety of ways, including distance education from well-respected engineering institutions, the use of in-house educational programs within engineering firms, agencies, and technical societies that are able to provide educational experiences that are documented to be equivalent in content, rigor, learning, and assessment to current engineering education. The American Society of Civil Engineers also recommends that those who teach engineers should predominantly be licensed.

The move forward to reform engineering education will require consistent and effective collaboration among members of academia, industry, and professional engineering organizations. If engineers are going to be adequately prepared to work in the knowledge-based 21st-century society, engineering education must undergo immediate reform—reform institutionalizing a master's degree program that provides the skill set needed by engineers to work in the 21st-century world.

As Claudio da Rocha Brito and Melany da Rocha Brito put it, "As engineering is the agent of progress and so the agent of transformation of human life, it is time not only to meditate but to change, and to do effectively something to make it better" (da Rocha Brito and da Rocha Brito, undated).

8

Proposal for a Master of Professional Engineering Management

THE NEED FOR WELL-QUALIFIED ENGINEERS COULD not be more critical—the world is faced with complex problems that affect the quality of life of everyone everywhere. And, as I note in previous chapters, businesses around the globe are crying out for more "well-rounded" engineers who can take on leadership roles. At the same time, paradoxically, the public's perception of the engineering profession is on a downward spiral, as is the enrollment of young people in engineering schools.

Now is the time to take action and enhance the engineer's skills. Although the need for enhanced professional skills and additional coursework is recognized by engineering educators in the United States and abroad, the work done to achieve these goals has been largely piecemeal and lacking in a total-system approach.

This chapter surveys what engineering schools outside and inside the United States are doing to meet the challenge of educating the engineer of the future. While several schools have begun offering courses that will meet the new ABET criteria and/or assist in the philosophy of the body of knowledge proposed by ASCE, no schools surveyed offer what the public and businesses are seeking: a program that gives an engineer the necessary skills to work in the 21st century. Thus, I propose a master's degree program in professional engineering management that can provide the training and skills required for engineers to be competitive and to succeed. I discuss specific details of the skill components in previous chapters. In this chapter, I propose a program for teaching those skills at universities, addressing the needs not only of the university, but of practicing engineers and their employers.

The concept of additional skill sets is not new. In 1945, as part of Cornell University's College of Engineering postwar planning, the university adopted a five-year undergraduate curriculum, which was founded on the belief that conventional four-year programs did not adequately prepare students for the practice of engineering. This program was initially phased in as an optional program but became mandatory during the 1947–1948 academic year. This bachelor of science program continued until the mid-1960s, when the university's Engineering Policy Committee raised concerns about the five-year undergraduate curriculum, recognizing that many changes had occurred since the implementation of the five-year bachelor's curriculum, among them a marked increase in the ratio of graduate students to undergraduate students in engineering programs; a national trend to award master's degrees for five-year programs of study in engineering; better preparation of students entering the undergraduate programs; the demand by students for the opportunity to prepare themselves for careers in new areas of engineering; and the increasing trend at Cornell to include graduate-style courses in the five-year undergraduate program.

In 1964, Cornell's Engineering Policy Committee issued a report (Cornell 1964) in which it recommended that the College of Engineering offer two coordinated degree programs: a four-year sequence leading to an undesignated bachelor's degree in engineering and a five-year sequence structured as the four-year sequence but that would have as its core courses that were more specifically selected to permit a student to qualify for a master of engineering degree in a designated field in addition to a bachelor's degree. The committee based its recommendation on two changes within engineering practice: a large increase in the complexity and diversity of engineering technology and the growing extent to which the practice cut across the boundaries of traditional engineering fields.

In May 1965, the College of Engineering faculty reaffirmed the retention of five-year programs but recommended that the appropriate degree for students satisfactorily completing this program would be the master of engineering degree. The faculty also voted to establish four-year undergraduate programs, recommending that the appropriate degree for students satisfactorily completing one of the programs be the undesignated bachelor of engineering degree. The addition of the master's program offered the four-year baccalaureate graduate the option of pursuing the professionally oriented master of engineering degree or the research-oriented master of science/doctoral program at Cornell (McGuire and Berth 1965).

More than 30 years passed, however, before the need for a broader engineering education again became the subject of serious reflection by a serious engineering body. In October 2001, the American Society of Civil Engineers' Board of Direction adopted Policy Statement 465, "Academic Prerequisites for Licensure and Professional Practice," which states that the society "supports the concept of the master's degree or equivalent as a prerequisite for licensure and the practice of civil engineering at the professional level." At the same time, ASCE established its Task Committee on Academic Prerequisites for Professional Practice to "develop, organize, and execute a detailed plan for full realization of Policy Statement 465." This committee subsequently established the Body of Knowledge Committee, which was tasked with defining the body of knowledge that would be required for professional practice. In 2004, the committee published *Civil Engineering Body of Knowledge for the 21st Century: Preparing the Civil Engineer for the Future*, calling for substantial reform in engineering education. The report notes,

> Civil engineers are expected to simultaneously possess broader capability and greater specialized technical competence than was required of previous generations—a nearly impossible challenge with fewer required college credits. Students earn at least 20 fewer credits than did their counterparts in the 1920s. While they take comparable proportions of mathematics, science, and general education, today's students complete, on average, 18 fewer credits of engineering topics. That is a whole semester less of technical education at a time when, by almost universal agreement, the complexity of the modern engineering project escalates. How can tomorrow's civil engineers design safe, cost-effective projects, accounting for greater complexity and uncertainty, with less formal education? (ASCE 2004, 13)

During its 2006 annual meeting, the National Council of Examiners for Engineering and Surveying—a national nonprofit group comprising engineering and surveying licensing boards representing all U.S. states and territories—voted to modify section 130.10 of its model law, which was established to provide guidance to states as they craft legislation pertaining to licensure. The change—which will become effective in 2015—calls for education beyond a bachelor of science in a program accredited by the Engineering Accreditation Commission of ABET, Inc. Students will be required to earn either a master's in engineering or thirty additional credits in upper-level undergraduate or graduate courses from approved course providers.

Indeed, the need for engineering education reform has captured the attention of many thoughtful individuals over the course of the past decade, and the consensus among these individuals is that universities must very quickly embrace the concept that the 21st-century engineer requires not only a solid command of engineering, but also a skill set that includes effective communication, a thorough understanding of globalization, team building abilities, and the ability to anticipate the needs of a workforce working both nationally and internationally. And yet despite the widely recognized need for engineering education reform, not a single college or university is offering a program comparable to the master of professional engineering management degree program I outline in this chapter—a master's program that focuses on the professional skills necessary in today's global, knowledge-based society. Several universities are offering courses within their engineering programs that do emphasize skills other than essential technical skills and that are similar to some of the courses I propose. But these courses are being offered on a piecemeal basis. What is urgently needed is an overhaul of the master's degree programs in engineering management. These programs must acknowledge the new business landscape and, consequently, align their courses to providing training in the skills required by this new business world.

The Current Engineering Education Landscape

Programs outside the United States

Several years ago, a number of universities in India recognized the fact that if India's engineers held any realistic expectation of gaining an equal footing within the upper tiers of the world marketplace,

they would have to be more broadly educated than they had been traditionally. Thus, India made a conscious decision to focus its engineering education in such a way as to optimize the commercialization of
technical know-how. The areas deemed to be of particular importance
were those of finance, organization, management, and marketing. As
a result, many Indian engineering graduates now hold senior positions
in other fields—among them business, consulting, and finance.

The French are also actively involved in reforming their engineering educational system to meet the demands of the 21st century and are
seeking to combine formal technological training with an education in
the liberal arts. Additionally, they are investigating ways to emphasize
a global perspective and hope to integrate the French system with the
European Union's educational system. The French are convinced that
these efforts will result in a major reform of higher education curricula
that will better position them in the global marketplace.

The Helsinki University of Technology offers a degree in construction economics and management that addresses the managerial,
organizational, administrative, economic, and legal issues inherent in
the construction industry as a whole. The courses offer instruction
in construction economics and management, construction contracting, advanced construction project management, management of construction contractors, management of constructors and internationalization, Finnish cases in the international construction business, and
strategies in international construction (Helsinki University of Technology 2003).

The School of Civil and Environmental Engineering at Nan-
yang Technological University in Singapore offers a graduate diploma
in construction management. Those who graduate may apply for a
master of science in international construction management. Some of
the courses offered in that program include project financing, international construction and marketing, construction management, techniques of project planning and control, and value engineering and
managing quality.

Temple University in Japan offers a two-day course in practical
project management, which is actually the fourth module in a series
offered as part of an open enrollment global management program
course. Those individuals who complete eight or more modules earn
certification in international business management. Temple also offers
courses in the professional skills area that include civil engineering construction administration, engineering project management,

construction financial management, construction industry business, and strategic corporate management in engineering construction.

Kochi University of Technology in Japan has incorporated project management throughout its undergraduate and graduate engineering curricula, and in 2006 was awarded a Center of Excellence grant from the Japanese government to further enhance its engineering curricula through the studies of project management, thus assisting its graduates in becoming more competitive in the global marketplace.

The Kanazawa Institute of Technology offers a course entitled Society and the Engineer, which focuses on ethics. The course examines three cases well known in Japan involving nuclear energy development: the 1995 Monju fire, the 1997 Bituminization Demonstration Facility fire and explosion, and the 1999 criticality accident at JCO, a subsidiary of the Sumitomo Metal Mining Company.

The Tokyo Institute of Technology has developed a graduate program in its Department of International Development Engineering that is offered in both Japanese and English. The courses are taught in collaboration with the Japan International Cooperation Agency and are designed to solve problems within the arena of international development, placing particular emphasis on problems in underdeveloped nations. Courses taught here are also similar to those being proposed as components of the master of professional engineering management and include courses in the principles of international development projects, exercises in international development projects, writing and analytical skills, sustainable development and integrated management approaches, and the principles of international coexistence.

The University of Newcastle in Australia offers a master of engineering management, which is described by the university as follows:

> Engineering management—the art and science of planning, organising, allocating resources, and directing and controlling activities that have a technical component—is rapidly becoming recognised as a professional discipline. Engineering managers are distinguished from other managers by the fact that they possess both the ability to apply engineering principles and skills in organising and leading people with technical and non-technical priorities. The Master of Engineering Management program is designed to:
> - Assist in the transition to a softer spectrum of skills offering courses covering theories and applications of motivational needs, management styles, interpersonal skills, and systems approach;
> - Develop leadership and communication potential;

- Strengthen engineering skills by offering unique and cutting-edge engineering courses covering a range of areas of interest including mechanical, civil, chemical, electrical, and software engineering. (University of Newcastle 2005)

Courses offered by the University of Newcastle are similar to those that will be proposed in this chapter and include engineering and project management, total quality management, teamwork and leadership, political management and decision making, accounting and financial management, management and organizational behavior, risk management, and sustainability for engineers and scientists.

The Hong Kong University of Science and Technology offers several master's degree programs—on campus, online, and in various combinations of the two. Courses offered by the university that are similar to the master of professional engineering management include construction practice and communication; construction financial management; construction project delivery; and engineering risk, reliability, and decision making.

Programs at U.S. Universities

Recognizing that professional skills are requisite for 21st-century engineering practice, several universities in the United States have begun to shape their informational literature to underscore their emphasis on professional skills. For example, several years ago, the Department of Mechanical Engineering at Carnegie Mellon University recognized the need to instill within its students the understanding that engineering practice today entails not just technical skills but interpersonal and intrapersonal skills. To that end, the department set up industry roundtables and advisory boards that provide students with useful interaction and feedback. By involving industry in this program, Carnegie Mellon is moving to ensure that students acquire the professional skills that are essential to the 21st-century leader.

The Department of Civil and Environmental Engineering at the University of California at Berkeley offers a graduate degree in engineering and project management that covers several of the skill sets advocated in this chapter, with teaching and research centered around the following areas:

- project and corporate management;
- engineering management;

- construction management;
- project-based production systems;
- lean construction;
- risk assessment and management;
- design–construction–operation–maintenance integration;
- green design, engineering and management;
- management of technology;
- project finance; and
- construction disputes and claims resolution (Berkeley 2007)

The University of Southern California's Viterbi School of Engineering's Distance Education Network (DEN) offers online courses and degree programs specifically tailored to the engineering professional. The admission requirements and curriculum for on-campus and DEN students are identical. Two of its degree programs—the master of science in engineering management and the master in construction management—offer many of the professional skills courses that are outlined for the master of professional engineering management, but these courses are not required by both programs; instead they are distributed between the two programs. They include oral communication in business; business communication management; negotiation and persuasion; management consulting; construction accounting and finance; accounting concepts and financial reporting; project cost estimating, control, planning, and scheduling; finance fundamentals; construction law; general construction estimating; international construction and engineering; quality management for engineers; management of engineering teams; the political process in systems architecture design; communicating in the working world—managing diversity and conflict; international business communication; and management for engineers.

The University of Washington, in Seattle, offers a degree in architecture and urban planning construction management that includes several courses similar to those designed for the proposed master's program, such as an introduction to the construction industry, construction accounting, construction contract documents, construction estimating, project planning and control, competitive business presentations, project management, construction law, construction firm management, and managing international projects.

Washington University in St. Louis offers a master of construction management and construction engineering, which includes a number of courses comparable to those being suggested here for the proposed master's program. Some of these are construction cost engineering, construction cost estimating, legal aspects of construction, construction risk management, construction management of public projects, construction project planning and scheduling, quality processes in construction management, and construction claims.

The Ohio State University offers a construction management program that includes courses in some of the subject areas suggested for inclusion in the proposed degree program, such as construction forensics, construction network analysis, deterministic construction estimating and pricing, construction contracts and claims, construction project administration, construction risk and decision analysis, and management accounting.

The University of Wisconsin at Madison offers a master of engineering in professional practice (MEPP), which is closely aligned with both my proposed master's degree and the attendant proposed method of education. The MEPP is an online degree that is designed to enable the practitioner to advance his or her career as an engineering leader by participating in the advanced educational program while remaining in active practice. The MEPP degree

> provides a blend of technical and management expertise that you can apply to your current position while preparing for engineering leadership roles. A better investment than an MBA for most engineers, this two-year program will improve your ability to: lead engineering projects and teams, communicate effectively as a leader, make effective engineering and business decisions. (University of Wisconsin 2007)

This online program enables an engineer to complete the master's degree within two years without interrupting his or her career. The degree consists of 26 graduate credits obtained by completing 10 courses as well as two week-long summer residencies at the university. The courses that most closely align with those of my proposed master's program include engineering economic analysis and management, technical project management, communicating technical information, international engineering strategies and operations, and quality engineering and quality management.

The George Washington University in Washington, D.C., offers a distance-mode master of science in project management, which is similar to my proposed master's program. The program integrates distance-mode learning with a week-long on-campus residency. Students are required to attend a one-day orientation session on campus to meet faculty members and learn how the distance-learning mode will be facilitated and how students will interact with the school. The class communication is mediated through Web-based course management software. Blackboard, a private Web shell, facilitates the course interaction, enabling instructors to publish course content and administrative information conveniently. Blackboard also provides virtual classrooms, e-mail, chat rooms, and discussion groups. The course is organized in such a way that students must take basic courses before taking advanced courses. Some of the courses offered as part of this master's program include an introduction to project and program management; risk management (advanced); planning and scheduling (advanced); project estimation and cost management (advanced); project management applications (advanced); organization, management, and leadership; conflict management; executive decision making; financial accounting; and financial management.

Villanova University offers a master's certificate in project management practices. Villanova says that an advanced program in project management is needed because "the increased globalization of the workplace, coupled with the ever-increasing influx of change, has caused a significant shift in the way business is conducted, projects are managed, and success is determined. Project management has come into the forefront of the skill sets necessary for success in this environment" (Villanova 2007). The courses offered in this certification program include understanding project management practices; project management performance management; issues in project risk management; communication, motivation, and leadership; issues in project quality management; intercultural communication and understanding; project plan development; and the project management process.

In reviewing the course offerings of these various universities, several areas of emphasis emerge: communication, project management, leadership, the engineer's role in society, and the importance of ethics. Clearly, these areas are being emphasized at the postbaccalaureate level—and universities have devised flexible programs that do not disrupt engineers' careers. Numerous universities now offer programs online that enable engineers to complete much of their course work by means of distance learning rather than by attending on-campus classes. If universities are to be successful in attracting the working professional, and if the engineering profession is to continue to develop, it will be necessary for universities—working in collaboration with industry—to undertake significant reform with respect to the way advanced engineering education is delivered.

Proposed Master of Professional Engineering Management

The degree that I am proposing—the master of professional engineering management, or PEM—is a 30-hour degree program that requires a written thesis, the subject of which is to be agreed upon by the student and the student's university adviser. This degree program could be completed within two years, but could also be extended over a three-year period, affording an engineer considerable flexibility in accommodating the demands of personal life and career. The degree title—master of professional engineering management—speaks to the acquisition of the professional skills necessary to manage and lead engineering teams and projects. The course work proposed for this master's program is designed to enhance an engineer's proficiency in management and communication capabilities—capabilities that are absolutely essential now.

This proposed master's program is designed for those students who are already at work in professional practice and do not wish to disrupt their practice to return to school, yet who fully understand the need to enhance their professional competence. This program integrates distance learning with on-campus seminars, but is designed so that the large majority of course work can be completed online. It is also founded on the assumption that specific firms will take a keen interest in such a program given the fact that it will enable their employees to advance professionally without disrupting their professional responsibilities to their firms. In fact, some of the courses proposed can be tailored to meet the needs of a specific firm while simultaneously satisfying overall academic requirements. I believe that such tailoring can serve to attract industry support to the program, encourage substantial numbers of engineers to enroll in the program, and enhance the quality of the program by means of an industry/academe interface. Those courses that I believe lend themselves to such tailoring—and that can actually be taught on a firm's own premises—are so noted.

The Proposed Courses

The intent of the master of professional engineering management program is to provide the consulting engineer with the skills necessary to be competitive and effective in the 21st-century marketplace. In an effort to deliver the information most applicable in the global workplace, I am recommending that the majority of the subject matter discussed in the preceding chapters of this book be incorporated into the actual course curriculum for this program. I must emphasize the point, however, that the course descriptions that follow serve merely as a guide; they are tailored for use in university course description catalogs, to provide the prospective student and/or employer with an overall understanding of the topics to be covered in the course, and to outline the minimum knowledge and/or skills that will have been acquired upon course completion.

The courses have been grouped into three segments: those designed for on-campus study, those designed to meet a firm's specific needs and requirements, and those designed for online study. The program calls for the student to make an initial campus visit to meet faculty members and other students and to gain an understanding of how the course program will be conducted, especially with respect to online studies. The program also requires students to participate in two one-week intensive on-campus sessions to study subjects that are best presented by a combination of university professors and industry practitioners. This program is designed to instill international competitiveness, and thus I recommend that all of the course work be conducted in the international business language of English and that students' theses be written in English as well.

On-Campus Course Work

PEM NETWORK SKILLS FOR DISTANCE LEARNERS

This course serves as the introductory on-campus course during which the student meets the faculty; it also affords the student the opportunity to use, explore, and gain confidence in all of the learning and collaboration tools utilized in completing the distance learning courses online. The course offers practical tips for efficient distance learning and provides opportunities to work through software/hardware/network configuration problems. The course also addresses the challenges of time management and balancing employer expectations, personal and family needs, and course responsibilities. Course topics include

- software setup, course communications, backup and trouble-shooting, and security considerations at home, at work, and while traveling;
- information management, including e-mail effectiveness, using online document management, introducing instant messaging, professional Web searching, and file management;
- learning at a distance, including learning in online discussion forums, chat rooms and effective Web conferencing; and
- file management skills, including understanding file types and file formats, sharing data between applications, and mastering word processing templates and styles.

(One credit)

PEM LEGAL ASPECTS OF ENGINEERING AND CONSTRUCTION

This course is presented as a one-week intensive lecture series held on campus. Most topics will be covered by expert practitioners who will share their personal experiences and by internationally respected legal advisers who deal exclusively with international construction projects and international arbitration. This course deals with the numerous ways in which the legal system affects or controls the construction process. The first half of the course will address the construction process from development of the contract through execution. Topics include basic legal doctrines, the consultant/client relationship, contractor selection, the construction process, professional practice problems, and claim resolution. The course will focus on various approaches to contracting, project delivery systems, contract law, and the relationships among the various parties involved in the construction process. The various principal contract relationships and responsibilities (owner–contractor, contractor–subcontractor, owner–architect/engineer, and owner–engineer–constructor) are explored, as are the problems and disputes that typically arise from these relationships. Additional course topics include legal aspects of contract documents, drawings, and specifications; bids and contract performance; labor laws; governmental administrative agencies; regulatory agencies; legal issues arising from design and construction services; and liability awareness. The course also examines the advanced project development process—business planning and preproject planning for engineering, procurement, and construction projects.

The second half of the course addresses construction claims and is designed to present the basic foundations of the construction claims process, beginning with an understanding of the contractual basis for construction claims and concluding with the final resolution of claims. The course includes a detailed survey of the various standard contracts used in the construction industry—those produced by the Engineers Joint Contract Documents Committee (EJCDC), International Federation of Consulting Engineers (FIDIC), the Engineering Advancement Association of Japan, the American Institute of Architects (AIA), American General Contractors (AGC), and so on—and the specific clauses that form the basis of claims; the recognition of claims; the contract notice requirements; and communicating the basis of claims, the pricing of claims, and various methods of resolving claims, including settlement and alternative dispute resolution. The course also presents the technical, legal, and business requirements for processing claims in the construction industry. Case studies of the most frequent causes of claims (project management decision, delay, disruption, acceleration, unforeseen site conditions, and changes) are discussed, as is the way in which international arbitration panels determine proof for causation, entitlement, and damages. The role of the expert witness is also briefly introduced. (Three credits)

PEM RISK MANAGEMENT

This course—an intensive one-week lecture series held on campus—is conducted by university professors and internationally respected experts in risk management. The practitioners focus on specific case studies in which risk management was employed and on the savings, return on investment, effective project management, and claims minimization evident as a direct result of risk management identification, assessment, and monitoring. The course identifies the various types of risk encountered in the construction industry and explores the basic principles of risk management practices. The course explores risk from different points of view (financier, senior management, project team, client, and so on) and focuses on the development of risk mitigation procedures and execution. Risk identification techniques are introduced, as are methods of quantifying risk and risk allocation. The steps in developing a risk management plan—including identifying, analyzing, mitigating, and monitoring project risks—

are discussed. Both simplistic matrix types of risk identification and probabilistic methods are explored, including reliability-based probabilistic risk analyses, Monte Carlo simulation, decision tree models, influence diagrams, and @RISK software modeling. Fundamentals of integrated risk assessment and risk-based decision analysis are emphasized. Through case studies and discussions, the student develops an understanding of the basic principles of risk assessment and management. Practical examples are used extensively to demonstrate the application of the risk identification and risk assessment methods. (Three credits)

Industry-Based Courses

The following courses may be offered either online or in a firm-tailored situation developed to address the specific techniques and methods used by a particular firm, thus further enhancing the ability of the engineer to be more effective and efficient in his or her own organization. A firm's case studies may be utilized as specific training tools. The course may be taught either by university professors at one of the firm's locations and/or via video-conferencing technology.

PEM PROJECT MANAGEMENT

This course provides an overview of the construction industry and practice and of the planning, design, and construction phases of civil engineering projects. The course includes an introduction to various organizational structures and their functions, work breakdown structure, and the setting up of the project team management and administrative functions, including the respective duties of the engineering manager, the project manager, and the construction manager. The essentials of project management practices and methods are discussed, including the basis of contracting, employment practices and labor relations, project financing and accounting, safety practices, quality management, lean construction, risk management, and various insurances, including owner-controlled insurance programs and contractor-obtained insurance. The course provides an introduction to project planning, scheduling, cost estimating, budgeting, cost accounting, and cost controls as well as an introduction to procurement, value engineering, and quality control/assurance. The course discusses the understanding of change orders, change management, project delay and acceleration, coordination, communication, project documentation and record keeping, partnering, safety, and ensuring the effective completion of a quality project on schedule, within budget, and within compliance of the contract specifications. (Three credits)

PEM COST ENGINEERING

This course examines cost engineering principles. The fundamentals reviewed focus on principles and techniques used to address problems involving cost estimating, cost control, profitability analysis, and the integration with planning and scheduling. The course introduces the formalized procedures, tools, and techniques used in developing the project estimate during the planning stages and in updating the estimate throughout the project life, and the tools and techniques used in monitoring, managing, and controlling the cost of the project— including the fundamentals of estimating and reviewing the details of costs associated with material, labor, equipment, overhead, and profit. The course will also focus on principles of interpreting financial information and performing engineering-related economic analyses. Course topics include

- financial principles, including the implications of basic accounting and cost systems to engineering and the interpretation of financial data, budgets, and accounting summaries, and
- costing systems and management control, including activity-based costing, pricing strategies, decision making, life-cycle cost analysis and models, budgeting, earned-value, and risk analysis.

(Two credits)

PEM PROJECT PLANNING AND SCHEDULING

This course is designed to provide an understanding of the planning and scheduling process and reviews

- the fundamentals of planning and scheduling information systems;
- the identification of project activities, activity duration estimation, and activity logic and sequence;
- activity coding;
- manpower and cost resource loading and allocation;
- activity constraints, and
- the identification of potential project milestones.

The course provides an understanding of the critical path method (CPM) theory, legal implications, and practice. CPM network-based scheduling methodologies including arrow-on-arrow, precedence, and probablistic evaluation review techniques are presented. An introduction to commercial project management and scheduling software is provided, including its application to a variety of construction problems and projects, such as critical path, project float, and activity float. Various techniques for analyzing delay on a project will be introduced. The emphasis is on resolving delay disputes as they arise during the project, but after-the-fact delay analyses are explored as is their acceptance by local legal jurisdictions and international arbitration panels. (Two credits)

PEM QUALITY MANAGEMENT

This course introduces the various theories of quality and the tools applied to quality practices/principles in the engineering and construction management processes. The course focuses on modern quality concepts, tools, and techniques used to develop, implement, and maintain systems for improving quality and productivity. Exploration of the use of quality management and planning tools will assist in defining quality problems and opportunities, implement measurable solutions, and foster team-based strategies for continuous improvement. The course topics include

- Total Quality Management;
- Kaizen/total quality management concepts and principles, basic problem-solving techniques and tools, and defining good process improvement projects and group processes;
- quality management systems (QMS), international quality standards (e.g., ISO 9000 and QS-9000), QMS documentation requirements and structures; and
- quality assurance.

The course addresses standardization of improved methods through proper documentation using ISO 9000 and QS-9000 standards, basic auditing techniques, and synergy and links between QMS and strategic quality planning. (Two credits)

Online Courses

These courses may be tailored to various online teaching methods, depending on the requirements of the university.

PEM COMMUNICATING TECHNICAL INFORMATION

This course focuses on communication skills for engineers. Both oral and written communication skills are addressed, and the course reviews techniques used in oral reporting, oral presentations, project team meetings, and management briefings and reviews listening techniques, interviewing skills, conference and committee leadership, and methods of written communication—including e-mail; daily, weekly, and monthly reports; proposals; and letters. The course includes a series of workshops and practical exercises in construction presentation skills, teamwork, and leadership. The course focuses on the ability to communicate effectively to various audiences and provides strategies for using technical information effectively. This course makes use of weekly Web conferences, forums on communication topics, and recommended reading. The weekly Web conferences assume the form of seminar discussions and include presentations and occasional guest lecturers. Course topics cover practical skills for advanced leadership, including:

- audience analyses and strategies for technical, nontechnical, and mixed audiences;
- teamwork;
- such business communications as e-mails, memos, and letters, making note of style, tone, diplomacy, and political correctness;
- major communication projects, including reports, proposals, and teleconferences;
- both live and online technical presentations; and
- such research sources as libraries, databases, and Web resources.

Applications for research papers, biographical sketches, engineering reports, and correspondence are emphasized. In addition to submitting a written report or proposal for grading, one of the course requirements is a videotaped oral technical presentation. Facilitation of a Web conference (usually 15 minutes in duration) is also required. (Three credits)

PEM THE ENGINEER'S RESPONSIBILITY TO SOCIETY
AND PROFESSIONAL ISSUES

This course examines the business issues that constitute the framework within which practitioners apply technology. The course provides an introduction to the profession of engineering and its practice, including social obligations and ethical challenges. Topics covered include professionalism and ethics and professional liability insurance. The course focuses on

- responsibility of oversight and engineering managers;
- the influence of quality of engineering and construction works;
- cover-ups and delays in informing others of accidents;
- work not performed to specification;
- the importance of professional work to society;
- learning from past experiences and the transfer of information from more experienced engineers;
- incompetent and inexperienced behavior by engineers;
- careless design;
- negligence in watching and observing construction;
- poor management and management ethics;
- trust and communication;
- the engineer's responsibility when observing unethical or life-threatening matters even when not in charge;
- sustainability and its consideration in both design and construction;
- global ethical considerations, such as corruption of officials;
- the importance of belonging to and participating in professional society activities;
- the importance of a professional code of ethics; and
- professional licensure.

The course reviews actual case studies of ethical and unethical behavior and the results and/or consequences and lessons to be learned from these cases. (Two credits)

PEM TEAMWORK AND LEADERSHIP

This course provides a series of tools that enhance organizational communication, motivation, team building, and leadership. Coaching and team building exercises are introduced. Change management, power, and influence are investigated. Integrative approaches to organizational concepts, management principles, philosophy, and the differences in theory in public and private organizations are discussed. Methods of making complex decisions, identifying criteria and alternatives, setting priorities, allocating resources, strategic planning, resolving conflict, and making group decisions will be explored. The course reviews various forms of leadership—individual, group, and organizational—and examines the concepts of leadership versus management, motivation, conflict and negotiation, and empowerment. Group dynamics and team building exercises are key aspects of this course, including exercises focusing on group decision making, motivation, leadership, performance measurement, team diversity, conflict, and integration. (Two credits)

PEM PUBLIC POLICY ANALYSIS AND DECISION MAKING

This course explores the intimate interaction between the political process and the engineering design and its processes. The importance of understanding how public policy is developed and implemented is explored as is the engineer's role in contributing to the public policy process. Methods and techniques of analyzing, developing, and evaluating public policies and programs are discussed with emphasis on benefit-cost and cost-effectiveness analysis and concepts of economic efficiency, equity, and distribution. The role of the engineer in the political process is explored as is the importance of communicating with politicians and running for public office. (Two credits)

PEM MANAGING INTERNATIONAL PROJECTS

This course focuses on the management of international construction projects. Emphasis is placed on understanding the different project delivery approaches and contracting, examining common problems associated with managing construction projects outside of the engineer's native country, identifying the risks involved, and discussing possible solutions. The course examines techniques that are utilized to enhance global business interaction and makes a comparative examination and analysis of global trends and regional variations of engineering concepts, standards, and practices. By means of case studies, lectures, and discussions, the course offers a perspective on how cultural differences among countries affect interpersonal and business relationships and actions. This course describes and analyzes multinational and national engineering operations, summarizing best practices. Comparative regional and national engineering professional practice procedures and methods are explored from different regions of the world. The challenges of working across cultures is explored, including

- national and regional cultures;
- business and corporate cultures;

- languages, both verbal and nonverbal; and
- special concerns, including economic and legal issues and social and political structures.

The course focuses on the need for better understanding of the international engineering and construction markets and of how to enter those markets, of the financial resources available, of international financial assistance entities, and of the basic legal considerations and risks involved in international projects. The course also explores the importance of sustainability in all international projects including policy and planning for sustainable development while critically examining the concept of sustainability as a process of social, organizational, and political development. (Three credits)

PEM UNDERSTANDING INTERCULTURAL COMMUNICATION AND DIVERSITY

As project management becomes more and more global, engineers have the opportunity to interact with business associates from many different countries and cultures. Background differences among the various parties involved, however, can pose communication challenges. This course enhances awareness and understanding of cultural differences—be they founded on ethnicity, religion, gender, education, or national origin. The course explores the intentional and unintentional effects of nonverbal communication and the potential damage it can cause to business relationships. The course identifies sources of stereotypes, and recognizes the impact of stereotypes on the job and suggests strategies for adapting to different communication styles. The course focuses on the critical importance of a diverse workforce and the challenges the engineering profession faces in attracting women to the profession as well as retaining them in the profession. Recommended techniques and methods for increasing the number of women working in the engineering profession and in construction-related companies will be presented. (Two credits)

Conclusions

If engineers are to be adequately prepared to work in a knowledge-based 21st-century society across all borders of the world, there must be an immediate reform in engineering education. What is required is a master's program that will impart the knowledge and skills required to work within the global economy—knowledge and skills that are not provided at the bachelor's or master's level at present. Innovative approaches must be explored to motivate the working engineer to reenter the academic environment. Such approaches include distance learning, cooperative education between academe and industry, and lectures or lecture series delivered by experts in various fields. Although the master's program I am proposing must be structured within a 30-hour credit requirement, those 30 hours may be completed in nontraditional ways.

The master of professional engineering management is designed to meet the needs of those who are already at work in professional practice—to provide them with the professional skills and knowledge they need to succeed in the 21st-century workplace without requiring them to place their careers on hold while they complete graduate school. This master's program will provide engineers with the knowledge and skills now required of engineering professionals—specifically, an understanding of globalization; of the importance of ethics and professionalism; of how to work effectively with diverse, multinational teams; and of public policy.

It will be these fundamental capacities that will enable the 21st-century engineer to work effectively and to succeed—admired and respected by the public and regarded by members of government as professionals whose services are to be engaged on the basis of qualification, not price.

References

About.com. (2007). "Women Truck Designers: The Women behind the Scenes." *4-Wheel Drive/Offroading,* <http://4wheeldrive.about.com/library/weekly/aa062101b.htm> (July 9, 2007).

Accreditation Board for Engineering and Technology (ABET). (2006). *Engineering Change: A Study of the Impact of EC2000, Report, and Appendix D: Engineering Criteria 2000,* ABET, Baltimore, Md.

American Association of Engineering Societies (AAES). (1994). *The Role of Engineering in Sustainable Development: Selected Readings and References for the Profession,* AAES and World Engineering Partnership for Sustainable Development, Washington, D.C.

American Electronics Association (AEA). (2007). *We Are Still Losing the Competitive Advantage: Now Is the Time to Act,* AEA, Washington, D.C.

American Institute of Architects (AIA). (2004). *Code of Ethics and Professional Conduct,* <http://www.aia.org/SiteObjects/files/codeofethics.pdf> (April 26, 2007).

American Society of Civil Engineers (ASCE). (2001). *Engineering the Future of Engineering,* Report of the Task Committee on the First Professional Degree, ASCE, Reston, Va.

———. (2004). *Civil Engineering Body of Knowledge for the 21st Century: Preparing the Civil Engineer for the Future,* ASCE, Reston, Va.

———. (2005). *Raise the Bar* 2(3, August).

————. (2007a). *Official Register,* ASCE, Reston, Va.

————. (2007b). *The Vision for Civil Engineering in 2025* (draft), ASCE, Reston, Va.

ASCE News. (2004). "ASCE Releases Report on Body of Knowledge for 21st-Century Engineers." 29(3, March).

Asian Academies of Engineering. (2004). *Asian Engineers' Guidelines of Ethics,* adopted by the Chinese Academy of Engineering, the Engineering Academy of Japan, and the National Academy of Engineering in Korea, <http://www.eaj.or.jp/openevent/declaration/declaration_e.pdf> (April 26, 2007).

Augustine, Norman R. (1994). "L.A. Engineers." *The Bridge,* Fall.

————. (1998). Comments at AIAA Summer Meeting, June 16, <http://www.ieeeusa.org/policy/guide/quotes.html> (July 9, 2007).

Axelrod, Alan. (1999). *Patton on Leadership: Strategic Lessons for Corporate Warfare,* Prentice Hall, New Jersey.

Baum, R. (1980). *Ethics and Engineering Curricula,* Institute of Society, Ethics and Life Sciences, New York.

Berkeley. (2007). "Degree Programs." University of California, Berkeley, Department of Civil and Environmental Engineering, Engineering & Project Management Program, <http://www.ce.berkeley.edu/epm/degrees/programs.html> (August 12, 2007).

Bordogna, Joseph. (2003). Address to the Meeting of the Engineering Workforce Commission, Washington, D.C., October 29, 2003.

————. (2004). "Diversity in the Professions: A New Challenge for Success." Presentation to ABET Annual Meeting Session, Nashville, Tenn., October 29. <http://www.nsf.gov/news/speeches/bordogna/04/jb041029_abet.jsp> (July 9, 2007).

Brundtland, Gro Harlem. (1987). *Our Common Future: The Brundtland Report,* Oxford University Press for United Nations World Commission on Environment and Development, Oxford, U.K.

Bush, Vannevar. (1945). *Science: The Endless Frontier,* A Report to the President by Vannevar Bush, Director of the Office of Scientific Research and Development. U.S. Government Printing Office, Washington, D.C.

Carnegie Endowment for International Peace (CEIP). (2007). *What Is Globalization?* <http://www.globalization101.org/What_is_Globalization.html> (April 26, 2007).

Casartelli, Giovanni. (2001). *Consulting Services Manual: A Comprehensive Guide to Selection of Consultants,* World Bank, Washington, D.C.

Catalyst, Inc. (1999). *Women Scientists in Industry: A Winning Formula for Companies,* Catalyst, New York.

————. (2003). *Women in U.S. Corporate Leadership: 2003,* Catalyst, New York.

————. (2004). *The Bottom Line: Connecting Corporate Performance and Gender Diversity*, Catalyst, New York.

Committee on Prospering in the Global Economy in the Twenty-First Century (Committee on Prospering). (2007). *Rising above the Gathering Storm: Energizing and Employing America for a Brighter Economic Future*, National Academies Press, Washington, D.C.

Congressional Commission on the Advancement of Women and Minorities in Science, Engineering and Technology Development (CAWMSET). (2000). *Land of Plenty: Diversity as America's Competitive Edge in Science, Engineering, and Technology*, <http://www.nsf.gov/pubs/2000/cawmset0409/cawmset_0409.pdf> (May 1, 2007).

Cornell Chronicle. (2001). "Engineering Academy's President Will Talk on Diversity, Computing Growth." 32(29), April 5, 2001, 7.

Cornell University. (1964). *Proposed Modification of the Degree Program in the College of Engineering*, Report by Engineering Policy Committee, Cornell University, Ithaca, N.Y., April.

Cottrell, David. (2002). *Monday Morning Leadership: 8 Mentoring Sessions You Can't Afford to Miss*, Cornerstone Leadership Institute, Dallas, Texas.

da Rocha Brito, Claudio, and da Rocha Brito, Melany M.C.T. (undated). "A New Engineer for a New Global Market." Department of Applied Sciences and Mathematics, University Center of Lusíada, Brazil.

Elkington, John. (1998). *Cannibals with Forks: The Triple Bottom Line of 21st Century Business*, New Society Publishers, Stony Creek, Conn.

Fleddermann, Charles B. (2004). *Engineering Ethics*, Pearson Education, Upper Saddle River, N.J.

Friedman, Thomas L. (2005). *The World Is Flat: A Brief History of the Twenty-First Century*, Farrar, Straus and Giroux, New York.

Harris, Charles E., Jr., Pritchard, Michael S., and Rabins, Michael J. (2005). *Engineering Ethics: Concepts and Cases*, 3rd ed., Thompson/Wadsworth, Belmont, Calif.

Harris Interactive. (2004). "Doctors, Scientists, Firemen, Teachers and Military Officers Top List as 'Most Prestigious Occupations.'" *Harris Poll* 65(15 September), <http://www.harrisinteractive.com/harris_poll/index.asp?PID=494> (July 9, 2007).

Hatch, Sybil E. (2006). *Changing Our World: True Stories of Women Engineers*, ASCE Press, Reston, Va.

Hawkins, John N., and Cummings, William K. (eds.). (2000). *Transnational Competence: Rethinking the U.S.–Japan Educational Relationship*, SUNY Press, Albany, New York.

Helsinki University of Technology (HUT). (2007). "Studying Construction Economics and Management in HUT (CEM)," <http://www.tkk.fi/Yksikot/Rakentamistalous/Opetus/education.htm> (July 9, 2007).

Hoffman, Edward. (2005). "Message Relay, Project Leadership: Focus, Adapt, Trust." *Leadership in Project Management Annual 2005*, 1, 22.

Hoover, Herbert. (1961). "The Profession of Engineering." *The Memoirs of Herbert Hoover*, Vol. 1, Macmillan, New York.

Hope College. (2000). *Report of the Hope College Conference on Designing the Undergraduate Curriculum in Communication*, <http://www.hope.edu/academic/communication/ecc/report.html> (April 26, 2007).

Huang, Gary, Taddese, Nebiyu, and Walter, Elizabeth. (2000). *Entry and Persistence of Women and Minorities in College Science and Engineering Education*, Research and Development Report NCES 2000–601, National Center for Education Statistics, U.S. Department of Education, Washington, D.C.

Institute of Electrical and Electronics Engineers, Inc. (IEEE). (2005). "How to Communicate with Members of Congress: The Basics." *Engineers Guide to Influencing Public Policy*, <http://www.ieeeusa.org/policy/guide/basics.html> (May 2, 2007).

Institute of International Education (IIE). (1997). *Towards Transnational Competence—Rethinking International Education: A U.S.-Japan Case Study*, IIE, New York, N.Y.

Institution of Civil Engineers (ICE). (2004). *Code of Professional Conduct*, <http://www.ice.org.uk/downloads/CODE%20OF%20PROFESSIONAL%20CONDUCT.doc> (April 26, 2007).

International Federation of Consulting Engineers (FIDIC). (2004). *Conflict of Interest*, FIDIC Policy Statement 21, <http://www1.fidic.org/about/statement21.asp> (April 26, 2007).

Jackson, Shirley A. (2003). *Envisioning a 21st Century Science and Engineering Workforce for the United States: Tasks for University, Industry, and Government*, National Academies Press, Washington, D.C.

Japan Society of Civil Engineers (JSCE). (2004). Roundtable discussion on women in civil engineering, June 2, 2004, Tokyo, Japan.

Kennedy, Theodore C. (2006). "The 'Value-Added' Approach to Engineering Education: An Industry Perspective." *The Bridge*, 36(2), 14–16.

Kenney, Martin, and Dossani, Rafiq. (2005). "Offshoring and the Future of U.S. Engineering: An Overview." *The Bridge*, 35(3), 5–12.

Kydd, Peter. (2006). "A Toolkit for Sustainability." *Notes*, Parsons Brinckerhoff magazine, December, 11.

Lattuca, Lisa R., Terenzini, Patrick T., Volkwein, J. Fredericks, and Peterson, George D. (2006). "The Changing Face of Engineering Education." *The Bridge*, 36(2), 5–13.

Mackay, Hugh. (1994). *Why Don't People Listen? Solving the Communication Problem*, Pan Australia, Sydney.

Mantell, Murray I. (1964). *Ethics and Professionalism in Engineering*, Macmillan, New York.

Martin, M., and Schinzinger, R. (2004). *Ethics in Engineering*, 4th ed., McGraw-Hill, New York.

Maryland Higher Education Commission. (1999). *Standards for General Education Speech Communication Courses*, <http://mhec.stac.usmd. edu/app-spc.html> (April 26, 2007).

Maxwell, John. (2003). "Appraise and Improve Your Leadership Skills." *Atlanta Business Chronicle*, February 21.

Maynard, M. (1999). "Ford Windstar's Designing Women." *USA Today*, November 12.

McGuire, W., and Berth, D. (1965). "Cornell's New Engineering Degree Program." *The Cornell Engineer* (January).

Metcalfe, Jenni. (2002). "Communication Planning for Scientists and Engineers," *Workbook on Communication and Management and Media Skills for Scientists & Engineers*, Foundation for Education, Science and Technology.

National Academy of Engineering (NAE). (2002). *Raising Public Awareness of Engineering*, Lance A. Davis and Robin D. Gibbin, eds., National Academies Press, Washington, D.C.

———. (2004). *The Engineer of 2020: Visions of Engineering in the New Century*, National Academies Press, Washington, D.C.

———. (2005). *Educating the Engineer of 2020: Adapting Engineering Education to the New Century*, National Academies Press, Washington, D.C.

National Council of Examiners for Engineering and Surveying (NCEES). (2007). *Model Law*. (September 2006). NCEES, Clemson, South Carolina, <http://www.ncees.org/introduction/about_ncees/ncees_model_law. pdf> (July 9, 2007).

National Research Council (NRC). (1986). *Engineering Education and Practice in the United States: Engineering Undergraduate Education*, Panel on Undergraduate Education, Committee on the Education and Utilization of the Engineer, Commission on Engineering, National Academy Press, Washington, D.C.

National Science Board (NSB). (2006). *Science and Engineering Indicators 2006*, Vol. 1, National Science Foundation, Arlington, Va.

National Science Foundation. (2003). *New Formulas for America's Workforce: Girls in Science and Engineering*, National Science Foundation, Arlington, Va.

———. (2004). *Women, Minorities, and Persons with Disabilities in Science and Engineering 2004*, NSF 04-317, Arlington, Va.

Nichols, Ralph G., and Stevens, Leonard A. (1957). "Listening to People." *Harvard Business Review*, September 1, 85.

Nielsen, Kris R. (2006). "Risk Management Lessons from Six Continents." *Journal of Management in Engineering*, 22(2), 61–67.

Pear, Robert. (2005). "U.S. Minorities Are Becoming the Majority." *International Herald Tribune*, August 13.

Perkins, Dennis N. T., Holtman, Margaret P., Kessler, Paul R., and McCarthy, Catherine. (2000). *Leading at the Edge: Leadership Lessons from the Extraordinary Saga of Shackleton's Antarctic Experience*, AMACOM, New York.

Peterson, Ralph R. (2005). "Twenty-First Century Leadership Challenges." *Engineering Times*, National Society of Professional Engineers, May.

Pletta, Dan Henry. (1984). *The Engineering Profession: Its Heritage and Its Emerging Public Purpose*, University Press of America, Lanham, Md.

Project Management Institute (PMI). (2004). *Guide to the Project Management Body of Knowledge*, 3rd ed., ANSI/PMI 99-001-2004, Project Management Institute, Newtown Square, Pa.

Ramaley, Judith. (2004). "Engineering as the Practical Expression of Liberal Education." Distinguished Lecture, American Society for Engineering Education meeting, Salt Lake City, Utah, June 23.

Silberglitt, Richard, Antón, Philip S., Howell, David R., and Wong, Anny. (2006). *The Global Technology Revolution 2020, In-Depth Analyses: Bio/Nano/Materials/Information Trends, Drivers, Barriers, and Social Implications. Executive Summary*. Report TR-303-NIC, RAND Corporation, Santa Monica, Calif.

Society of Petroleum Engineers (SPE). (2006). "Validating Engineering Competence." *Journal of Petroleum Technology* 58(7), <http://www.spe.org/spe-app/spe/jpt/2006/07/SPE_news_jul06.htm> (July 9, 2007).

Strong, Brent. (2003). "Beat Back the Nerd and Awaken Your Inner Leader—Why Engineers Should Read Shakespeare." *Composites Fabrication*, February.

Sununu, John. (2003). *John Sununu Talks about the Role of Engineers in Public Policy. Purdue News*, March 24.

United Nations. (2004). *World Urbanization Prospects: The 2003 Revision*, Population Division, Department of Economic and Social Affairs, United Nations, New York.

———. (2007). *World Population Prospects: The 2006 Revision Population Database*, <http://esa.un.org/unpp/> (July 9, 2007).

United Nations Commission on Environment and Development (UNCED). (2007). Chairman's Summary, Fifteenth Session of the Commission on Sustainable Development, May 11.

United Nations Educational, Scientific and Cultural Organization (UNESCO). (2000). "Science Agenda—Framework for Action." *World Conference on Science: Science for the Twenty-First Century: A New Commitment*, UNESCO, Paris, 476–485.

U.S. Army Corps of Engineers. (2003). "Communication Guide." *USACE 2012,* <http://www.hq.usace.army.mil/stakeholders/Communications.htm> (April 26, 2007).

University of Newcastle. (2005). "Master of Engineering Management." *Program Finder,* <http://www.newcastle.edu.au/program/11277.html> (July 9, 2007).

University of Wisconsin. (2007). "MEPP and Your Engineering Career." Master of Engineering in Professional Practice, <http://mepp.engr.wisc.edu/why/career.lasso> (July 9, 2007).

Villanova University. (2007). "The Masters Certificate in Project Management." Villanova Continuing Studies, <http://www3.villanova.edu/continuingstudies/mcpmp/> (July 9, 2007).

Washington Accord. (2006). <http://www.washingtonaccord.org/wash_accord_agreement.html> (April 26, 2007).

Weingardt, Richard G. (1995). "Engineers as Lawmakers." *Civil Engineering News,* November.

Wisely, William H. (1978). "Public Obligation and the Ethics System." *Preprint No. 3415,* American Society of Civil Engineers, New York.

World Bank. (2006). *Consulting Services Manual 2006,* World Bank, Washington, D.C.

World Economic Forum. (2004). *Partnering against Corruptions: Principles for Countering Bribery,* <http://www.weforum.org/pdf/paci/principles_short.pdf> (April 26, 2007).

Index

ABET, Inc. (Accreditation Board for Engineering and Technology), 7, 41*t*, 92, 94–96, 99, 102
accreditation standards, 93–96
See also licensure
American Association of Engineering Societies, 59–60, 78
American Association of Engineers, 93
American Electronics Association, 88
American General Contractors (AGC), 112
American Institute of Architects (AIA), 42–43, 112
American Society for Engineering Education, 87, 94–95
American Society of Civil Engineers (ASCE)
  body of knowledge project, 6–7, 97–98, 99, 101
  code of ethics and professionalism, 35, 39–40, 41*t*, 42–44
  education reform activities, 87, 93
  female leadership, 59
  licensure reforms, 6, 101

American Society of Civil Engineers (ASCE)—*continued*
  Policy Statement 465, 6, 101
  Summit on the Future of Civil Engineering of 2006, 50
American Society of Mechanical Engineers (ASME), 39–40, 41*t*, 59
Asian Academies of Engineering, 39–40
Asia-Pacific Economic Cooperation (APEC), 7–8, 22
Association of Finnish Civil Engineers (RIL), 48
Association of Japanese Consulting Engineers (AJCE), 22, 41*t*
Augustine, Norman R., 78–79

bachelor's degree level, 2, 5, 7
  accreditation challenges, 94–95
  five-year programs, 100
  four-year programs, 90–97
  undesignated bachelor of engineering degree, 101, 102
Bagehot, Walter, 71
Barton, Bruce, 71
Bell, Alexander Graham, 70–71

Big Dig project in Boston, 14
Blackboard, 108
body of knowledge for civil engineers, 6–9
  APEC's *Engineer Manual,* 7–8
  ASCE body of knowledge project, 6–7, 97–98, 99, 101
  ASCE Policy Statement 465, 6–7, 101
  business and managerial. *See* managerial skills
  communication skills. *See* communication skills
  leadership skills. *See* leadership skills
  NAE recommendations, 8–9
  public policy skills. *See* political skills
  SPE skills matrix, 8
Bordogna, Joseph, 54, 61–62
bridge accidents, 36, 65–66
Bush, Vannevar, 51–52

Canadian engineering, 36, 37
Carnegie Mellon University, 105
Carter, Jimmy, 64, 79
Catalyst, 56–57
certification. *See* licensure
Changing Our World: True Stories of Women Engineers (Hatch), 61
Channel Tunnel Rail Link, 14
Civil Engineering Body of Knowledge for the 21st Century (ASCE), 6–7, 97–98, 101
codes of ethics, 38–41
  See also ethics
Commission on Advancement of Women and Minorities in Science, Engineering, and Technological Development (CAWMSET), 52–53
commitment, 72

Communicating Technical Information course (PEM), 116
communication skills, 2, 3, 25–32, 92, 94
  education and training, 30–32
  e-mail and Internet use, 13–14, 26–27
  formal communication, 26
  informal communication, 26–27
  leadership skills, 67–71
  listening skills, 26
  principles of communication, 27–28
  project communications management, 28–30
confidence, 71–72
Consulting Engineers Association of India (CEAI), 41*t*
continuing education, 2
  See also master of professional engineering management degree
Cornell University College of Engineering, 100–101
Cost Engineering course (PEM), 114
Crossrail project in London, 14
cultural skills, 5–6, 92
Cummings, William, 5–6

da Rocha Brito, Claudio and Melany, 98
delivery/operation risk, 20
developed nations, 16
developing nations, 16–17
Dimitrov District, Russia, 79–81
Disney, Walt, 71
distance learning courses, 106, 107, 109
  Network Skills course (PEM), 110–11
  PEM's online courses, 115–19

About the
Author

Patricia D. Galloway, ph.d., p.e., is chief executive officer
of The Nielsen-Wurster Group, Inc., an international management consulting firm with offices throughout the United
States and Asia-Pacific. She is a licensed professional engineer in 11 U.S. states, Canada, and Australia, and a certi-
fied project management professional. She earned a bachelor's degree in civil engineering from Purdue University,
a master's in business administration from the New York
Institute of Technology, and a doctorate in infrastructure
systems engineering from Kochi University of Technology
in Japan.

An internationally recognized leader in civil engineering and construction, Galloway has extensive experience in
dispute resolution, expert witness testimony, risk management, and global engineering. She has worked with owners,
contractors, and engineering companies and has visited
more than 100 countries in a professional capacity. She has
lectured widely on topics such as leadership, globalization,
engineering education, and women in engineering.

In 2003, Galloway became the first woman to serve
as president of ASCE, and in 2006 she was appointed to the
National Science Board. She is a member of, among others,
ASCE, the National Academy of Construction, the Project
Management Institute, and the Pan American Academy of
Engineering, where she serves on the Board of Directors.